THE COMMONWEALTH AND INTERNATIONAL LIBRARY

MATHEMATICS DIVISION

ELEMENTARY VECTORS

SECOND EDITION

ELEMENTARY VECTORS

E. Œ. WOLSTENHOLME

SECOND EDITION

PERGAMON PRESS
OXFORD · NEW YORK · TORONTO
SYDNEY · BRAUNSCHWEIG

Pergamon Press Ltd., Headington Hill Hall, Oxford

Pergamon Press Inc., Maxwell House, Fairview Park, Elmsford, New York 10523

Pergamon of Canada Ltd., 207 Queen's Quay West, Toronto 1

Pergamon Press (Aust.) Pty. Ltd., 19a Boundary Street, Rushcutters Bay, N.S.W. 2011, Australia

Vieweg & Sohn GmbH, Burgplatz 1, Braunschweig

First published 1964

Revised and reprinted with corrections 1967

Second edition 1971

Library of Congress Catalog Card No. 73–141684

Printed in Great Britain by A. Wheaton & Co., Exeter

08 016569 9 (hard cover)

08 016570 2 (flexicover)

CONTENTS

PREFACE

THE aim of this book is to provide an introductory course in vector analysis which is both rigorous and elementary, and to demonstrate the elegance of vector methods in Geometry and Mechanics.

I should like to express here my gratitude to Dr. E. A. Maxwell for his help and encouragement in reading the original MS and to the staff of Pergamon Press for their help in the preparation of the MS for printing. I am also grateful to my colleague Miss D. W. Fielding for assistance in preparing the diagrams. I am indebted to the Senate of London University for permission to use examples from London B.Sc. papers, and to the Senate of Sheffield University for permission to use examples from B.Sc. and other Sheffield University papers.

An extra chapter, Chapter VII, has been included in this book, giving a parametric treatment of certain three-dimensional curves and surfaces including the helix. This has been done in responce to requests from teachers who wish to have a book covering the requirements of those G.C.E. A-level syllabuses which now include some work on vectors.

Note for American readers: In Chapters III, IV, V, the term "integration" should be regarded as synonymous with "anti-differentiation".

E. Œ. WOLSTENHOLME

CHAPTER I

§1.1. Real Numbers and Scalar Quantities

Any physical quantity which can be completely represented by a real number is known as a *scalar quantity*, or simply as a *scalar*. Thus a scalar quantity has magnitude, including the sense of being positive or negative, but no assigned position, and no assigned direction. Examples of scalars are *mass*, *energy*, *time*, *work*, *power*, *electrical resistance*, and *temperature*.

§1.2. Vector Quantities

Consider now a space Σ in which a point O has been arbitrarily chosen as an origin. Then any point A in Σ may be said to define both a *magnitude*, represented by the distance between O and A, and a *direction*, represented by the direction *from O to A*. Any quantity which can be completely represented by such a pair of points O and A is known as a *vector quantity*, or a *vector*. Thus if a vector is known to have a certain direction, and a certain magnitude a, an origin O may be chosen and through O a line OA may be drawn in the given direction and of a length to represent a; the vector is then completely represented by the *displacement* from O to A.

§1.3. Notation

The vector quantity represented by the pair of points O and A in §1.2 above is denoted by $\overline{OA}$ or $\mathbf{a}$. The number a represented by the distance OA is always positive and is known as the *modulus* of the vector quantity. The modulus may be written as $|\overline{OA}|$ or $|\mathbf{a}|$ or OA or a according to convenience in any particular context.

If $\overline{OA}$ represents the vector **a**, then $\overline{AO}$ is said to represent the vector $-$ **a**.

§1.4. Nomenclature

A vector quantity as defined in §1.2 has magnitude and direction but no assigned position in space, as the initial point O was arbitrarily chosen. Such a vector quantity is known as a *free vector.* When the term *vector* is used, it is assumed that it refers to a *free vector.*

If, however, the vector quantity has not only a specified magnitude and direction, but must be located in a specified line in the given direction, the vector quantity is known as a *line vector.*

If, on the other hand, instead of an arbitrarily-chosen origin O there is a specified point O which must be taken as origin, then only one point A is needed to complete the representation of this restricted vector quantity which is known as a *position vector*, or, more precisely, as the *position vector of A.*

If **a**, **b** are two free vectors, the expression "the plane of **a** and **b**" is understood to mean any plane in which can be drawn two lines, one parallel to **a** and one parallel to **b**. There is an infinite number of such planes, for through an arbitrary point O it is always possible to draw two lines OA, OB in the directions of **a**, **b** respectively, thus defining a plane OAB which conforms with the definition.

§1.5. Equivalence of Two Vectors

If **a** is a vector and P, R are two points in space, then points Q,S may be found such that $PQ = RS = a$, the displacement from P to Q is in the same direction as **a**, and the displacement from R to S is also in the same direction as **a**. The vectors $\overline{PQ}$ and $\overline{RS}$ are then said to be *equivalent vectors*. This may be written

$$\overline{PQ} = \overline{RS} = \mathbf{a}.$$

Similarly

$$\overline{QP} = \overline{SR} = -\mathbf{a}.$$

§1.6. Sum of Two Vectors

Assuming a vector to be completely represented by a displacement, suppose two vectors, **a**,**b** are represented by the displacements $\overline{FG}$ and $\overline{GH}$ respectively. Then the vector represented by the displacement $\overline{FH}$, which is equivalent to the displacement from F to G followed by the displacement from G to H, is defined to be the sum of the vectors **a** and **b**.

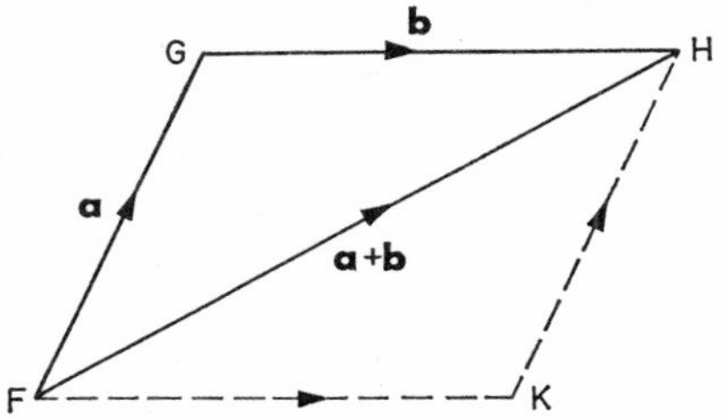

Fig. 1

This may be written

$$\overline{FH} = \overline{FG} + \overline{GH} = \mathbf{a} + \mathbf{b}.$$

If the parallelogram $FGHK$ is completed, the displacement from F to H can be seen to be equivalent also to the displacement from F to K followed by the displacement from K to H,

i.e. $$\overline{FH} = \overline{FK} + \overline{KH}.$$

But $$\overline{FK} = \overline{GH} = \mathbf{b}$$

and $$\overline{KH} = \overline{FG} = \mathbf{a},$$

hence $$\overline{FH} = \mathbf{b} + \mathbf{a}.$$

Thus $\mathbf{a} + \mathbf{b} = \mathbf{b} + \mathbf{a}$, and the commutative law for addition in scalar algebra is found to apply also to addition in vector algebra.

§1.7. Difference of Two Vectors

Suppose that **a** and **b** are two vectors, and, with the notation of §1.6, that points F, G, H are taken so that $\overline{FG} = \mathbf{a}$, $\overline{GH} = \mathbf{b}$; suppose further that HG is produced to H' so that $GH' = HG$. Then

$$\overline{H'G} = \overline{GH} = \mathbf{b}$$

$$\overline{GH'} = \overline{HG} = -\mathbf{b}.$$

The displacement from F to G followed by the displacement from G to H' is equivalent to the displacement from F to H',

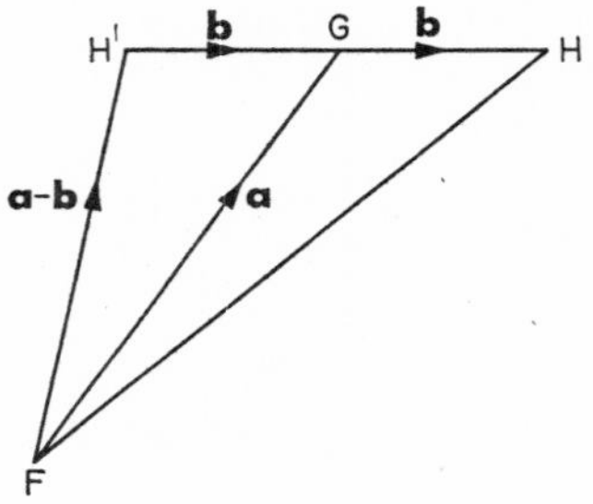

Fig. 2

i.e. $$\overline{FH'} = \overline{FG} + \overline{GH'},$$

i.e. $$\overline{FH'} = \mathbf{a} + (-\mathbf{b}). \tag{1}$$

Assuming as in real scalar algebra that $\mathbf{a} - \mathbf{b}$ is equivalent to $\mathbf{a} + (-\mathbf{b})$, equation (1) becomes

$$\overline{FH'} = \mathbf{a} - \mathbf{b}.$$

§1.8. Multiplication of a Vector by a Real Number

Suppose that **a** is a vector and that n is a real number. The result of multiplying **a** by n is defined to be the vector $n\mathbf{a}$ whose modulus

$n\mathbf{a}$ is n times the modulus of **a** and whose direction is the same as the direction of **a**.

§1.9. Sum of a Number of Vectors

Suppose that $\mathbf{a}_1, \mathbf{a}_2, \ldots, \mathbf{a}_n$ is a set of vectors whose sum is required. An arbitrary point O may be chosen, and then a point A_1 may be found such that $\overline{OA_1} = \mathbf{a}_1$. A point A_2 may then be found such that $\overline{A_1A_2} = \mathbf{a}_2$. Then by §1.6,

$$\overline{OA_2} = \overline{OA_1} + \overline{A_1A_2},$$

i.e.

$$\overline{OA_2} = \mathbf{a}_1 + \mathbf{a}_2.$$

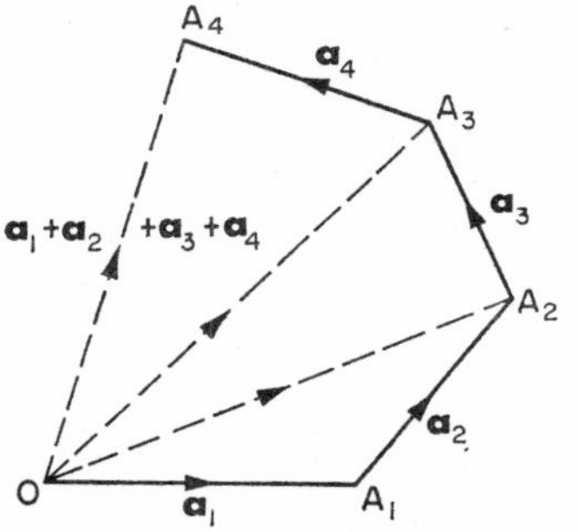

Fig. 3

A point A_3 may now be found such that $\overline{A_2A_3} = \mathbf{a}_3$.

Then

$$\overline{OA_3} = \overline{OA_2} + \overline{A_2A_3}$$

i.e.

$$\overline{OA_3} = \mathbf{a}_1 + \mathbf{a}_2 + \mathbf{a}_3,$$

and in general,

$$\overline{OA_n} = \mathbf{a}_1 + \mathbf{a}_2 + \ldots + \mathbf{a}_n.$$

If some of the vectors are to be subtracted, e.g. $\mathbf{a}_1 - \mathbf{a}_2 + \mathbf{a}_3 + \ldots$, the problem may be reduced to the process of addition by writing the expression in the form

$$\mathbf{a}_1 + (-\mathbf{a}_2) + \mathbf{a}_3 + \ldots$$

§1.10

THEOREM. *If* **a** *and* **b** *are two vectors represented by* $\overline{OA}$ *and* $\overline{OB}$ *and if C is a point in AB such that* $AC : CB = \mu{:}\lambda$, *where* λ, μ *are real numbers, then* $\lambda\mathbf{a} + \mu\mathbf{b} = (\lambda + \mu)\,\mathbf{c}$ *where* $\mathbf{c} = \overline{OC}$.

Since $$\overline{OA} = \overline{OC} + \overline{CA},$$

therefore $$\lambda\overline{OA} = \lambda\overline{OC} + \lambda\overline{CA}.$$

Similarly $$\overline{OB} = \overline{OC} + \overline{CB},$$

therefore $$\mu\overline{OB} = \mu\overline{OC} + \mu\overline{CB}.$$

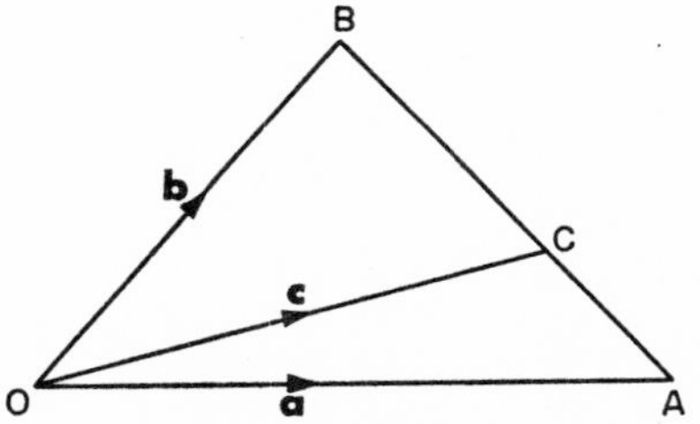

Fig. 4

Therefore $$\lambda\overline{OA} + \mu\overline{OB} = (\lambda + \mu)\overline{OC} + \lambda\overline{CA} + \mu\overline{CB}. \quad (1)$$

But $$AC{:}CB = \mu{:}\lambda,$$

i.e. $$\lambda AC = \mu CB.$$

Hence, having regard to sense,

$$\lambda\overline{CA} = -\,\mu\overline{CB}$$

i.e. $$\lambda\overline{CA} + \mu\overline{CB} = 0.$$

Therefore (1) now becomes

$$\lambda\overline{OA} + \mu\overline{OB} = (\lambda + \mu)\overline{OC},$$

i.e. $$\lambda\mathbf{a} + \mu\mathbf{b} = (\lambda + \mu)\mathbf{c}.$$

Example 1. *If ABCD is a quadrilateral in which H, K are the mid points of BC, AD respectively, show that* $\overline{AB} + \overline{DC} = 2\overline{KH}$.
Considering the quadrilateral *ABHK*,

$$\overline{AB} = \overline{AK} + \overline{KH} + \overline{HB};$$

considering the quadrilateral *DCHK*

$$\overline{DC} = \overline{DK} + \overline{KH} + \overline{HC}.$$

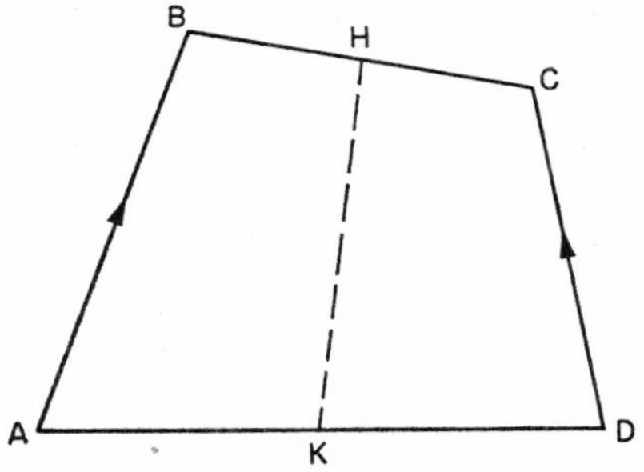

Fig. 5

Therefore $\overline{AB} + \overline{DC} = (\overline{AK} + \overline{DK}) + 2\,\overline{KH} + (\overline{HB} + \overline{HC})$.

But since *H*, *K* are the mid-points of *BC*, *AD*,

therefore $$BH = HC, \text{ and } AK = KD.$$

Therefore $$\overline{HB} = -\overline{HC}, \text{ and } \overline{AK} = -\overline{DK},$$

i.e. $$\overline{HB} + \overline{HC} = 0 \text{ and } \overline{AK} + \overline{DK} = 0,$$

therefore $\overline{AB} + \overline{DC} = 2\,\overline{KH}$.

Example 2. *ABCD is a square. A system of forces is completely represented by the lines AB, BC, AC, the directions of the forces being indicated by the order of the letters. Find the resultant of these forces and state where its line of action cuts AB.*

Let $\mathbf{P}_1$, $\mathbf{P}_2$, $\mathbf{P}_3$ be the three given forces where $\mathbf{P}_1 = \overline{AB}$, $\mathbf{P}_2 = \overline{BC}$, $\mathbf{P}_3 = \overline{AC}$. Let L, M, be the mid-points of AB, BC respectively.

Then
$$\mathbf{P}_1 + \mathbf{P}_3 = \overline{AB} + \overline{AC}$$
$$= 2\,\overline{AM}$$
$$= \mathbf{P}_4, \text{ say.}$$

Since the forces $\mathbf{P}_1$, $\mathbf{P}_3$ both act through A, then their resultant $\mathbf{P}_4$ is a force of magnitude $2\,AM$ acting in the line AM in the direction from A to M.

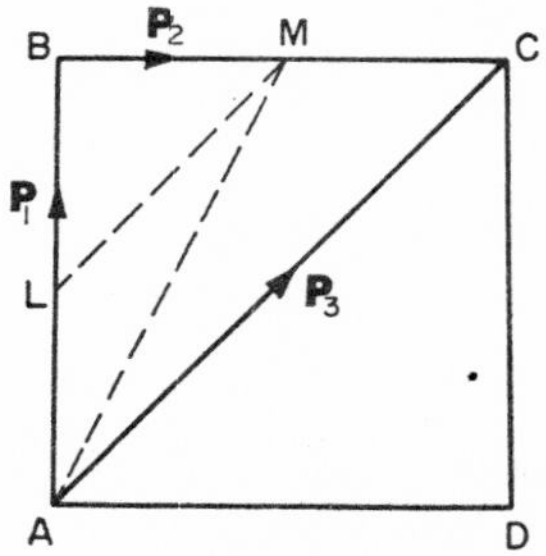

Fig. 6

Then
$$\mathbf{P}_1 + \mathbf{P}_2 + \mathbf{P}_3 = (\mathbf{P}_1 + \mathbf{P}_3) + \mathbf{P}_2$$
$$= \mathbf{P}_4 + \mathbf{P}_2$$
$$= 2\,\overline{AM} + 2\,\overline{MC}$$
$$= 2\overline{AC} \text{ in magnitude and direction.}$$

Since both $\mathbf{P}_2$, $\mathbf{P}_4$ act through M, their resultant also acts through M, and is of magnitude $2\ AC$ acting in the direction from L to M, i.e. the resultant of $\mathbf{P}_1$, $\mathbf{P}_2$, $\mathbf{P}_3$ is a force of magnitude $2\ AC$ acting through the mid point of AB in a direction parallel to, and in the same sense as, $\overline{AC}$.

N.B. *If AB is a straight line and n is a real number, a force* $\mathbf{P}$ *is said to be completely represented by nAB if the magnitude of* $\mathbf{P}$ *is nAB, and the line of action of* $\mathbf{P}$ *is AB in the direction $\overline{AB}$.*

Examples Ia

1. F, H, are the mid-points of the diagonals AC, BD of the quadrilateral $ABCD$. Prove that

$$\overline{BA} + \overline{BC} + \overline{DA} + \overline{DC} = 4\overline{HF}.$$

2. ABC is a triangle and O is any point in space. Prove that $\overline{OA} + \overline{OB} + \overline{OC} = \overline{OD} + \overline{OE} + \overline{OF}$ where D, E, F are the mid-points of BC, CA, AB respectively.

3. $VABC$ is a tetrahedron and X is the mid-point of BC. Prove that $\overline{VA} + \overline{AC} + \overline{VB} = 2\overline{VX}$.

4. $ABCD$ is a quadrilateral. Show that the resultant of forces represented completely by AB, BC, AD, DC, the directions being indicated by the order of the letters, is a force acting through the mid-point of BD, and represented by the vector $2\overline{AC}$.

5. ABC is a triangle and O is an arbitrary point. If G is the centroid of triangle ABC, show that $\overline{OA} + \overline{OB} + \overline{OC} = 3\overline{OG}$.

6. A, B, are two fixed points on the circumference of a circle and P is a variable point on the circumference of the same circle. Show that if the resultant of forces completely represented by $2PA$ and $3PB$, in the directions indicated by the order of the letters, is completely represented by PQ, then the locus of Q is a circle. Find the radius of this circle in terms of the radius of the circle APB.

7. XY is a fixed diameter of a circle and A is an arbitrary point on the circumference. If $\overline{AX} + \overline{AY} = \overline{AB}$ find the locus of B as A moves on the circle.

8. If AB, $A'B'$ are any two lines in space, show that

$$\overline{AA'} + \overline{BB'} = 2\overline{GG'}$$

where G, G' are the mid-points of AB, $A'B'$ respectively.

9. If A, B, C, D, E, F, are the angular points of a regular hexagon, prove that

$$\overline{AB} + \overline{AC} + \overline{AD} + \overline{AE} + \overline{AF} = 3\overline{AD}.$$

10. If AA', BB', CC', DD' are parallel edges of a parallelepiped, and AC' is a diagonal, show that

$$\overline{AB} + \overline{AC} + \overline{AD} + \overline{AA'} + \overline{AB'} + \overline{AC'} + \overline{AD'} = 4\overline{AC'}.$$

11. Forces represented by $\overline{AB}, \overline{BC}, \overline{AD}, \overline{DC}$ act at A. Show that their resultant is a force acting at A and represented by $2\overline{AC}$.

12. Prove that the resultant of the forces $\lambda\overline{OA}$ and $\mu\overline{OB}$ acting at O is $(\lambda + \mu)\overline{OC}$ also acting at O, where C is the point in AB such that $AC : CB = \mu : \lambda$.

Forces $\lambda\overline{BC}, \mu\overline{CA}, \nu\overline{BA}$ act along the sides of the triangle ABC. The resultant cuts BC, CA in P, Q respectively. Show that the resultant is

$$\nu^{-1}(\lambda + \nu)(\mu + \nu)\overline{PQ}.$$

§1.11. Unit Vector

Suppose **a** is a vector and suppose **e** is a unit vector in the direction of **a**, i.e. a vector whose direction is the same as that of **a** and whose modulus is unity. Then $\mathbf{a} = a\mathbf{e}$.

§1.12. Components of a Vector

Suppose $\overline{OP}$ represents the vector **P**. Let OA, OB, OC be three

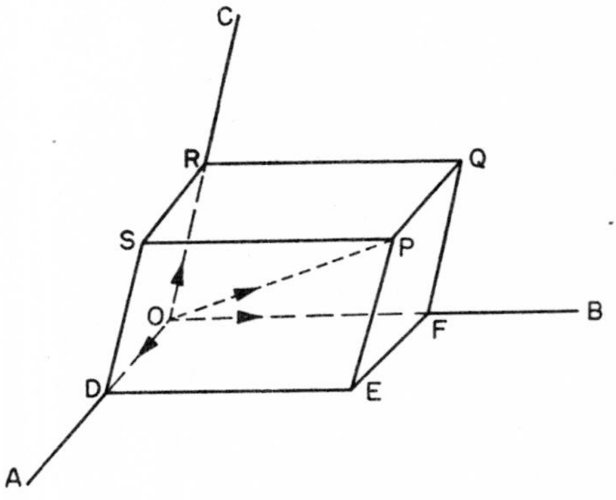

Fig. 7

non-coplanar lines through O. Complete the parallelepiped $ODEFRSPQ$, having its edges OD, OF, OR, along OA, OB, OC respectively. Then

$$\overline{OP} = \overline{OD} + \overline{DE} + \overline{EP},$$

i.e.

$$\overline{OP} = \overline{OD} + \overline{OF} + \overline{OR}. \tag{1}$$

If the vectors represented by $\overline{OD}$, $\overline{OF}$, $\overline{OR}$ are $\mathbf{P}_1$, $\mathbf{P}_2$, $\mathbf{P}_3$ respectively, then (1) can be written

$$\mathbf{P} = \mathbf{P}_1 + \mathbf{P}_2 + \mathbf{P}_3.$$

$\mathbf{P}_1$, $\mathbf{P}_2$, $\mathbf{P}_3$ are known as the *component vectors* of $\mathbf{P}$ along OA, OB, OC respectively.

It is usual to take the three directions OA, OB, OC, to be mutually perpendicular, and to conform in cyclic order to the right-handed screw rule: i.e. if a right-handed screw were placed with its

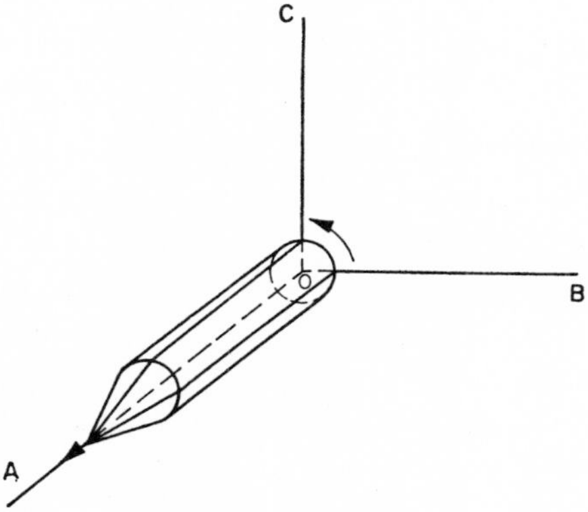

Fig. 8

axis along OA and turned in the positive sense so that the point travelled in the direction from O to A, a line fixed in the screw normal to its axis, would rotate in the sense which would bring it, in a right-angle turn, from the direction OB to the direction OC;

similarly if the screw were turned right-handedly so that its point travelled along *OB*, the fixed line normal to the axis would rotate through one right-angle from the direction of *OC* to the direction of *OA*; and if the point travelled along *OC*, the normal to the axis would rotate through one right angle from the direction of *OA* to the direction of *OB*.

Suppose *OX*, *OY*, *OZ* is such a right-handed system of mutually perpendicular axes, and suppose $\overline{OA}$ represents the vector **a**. Complete the rectangular parallelepiped *OPQRMNAL* having its edges *OP*, *OR*, *OM*, along *OX*, *OY*, *OZ* respectively. Suppose

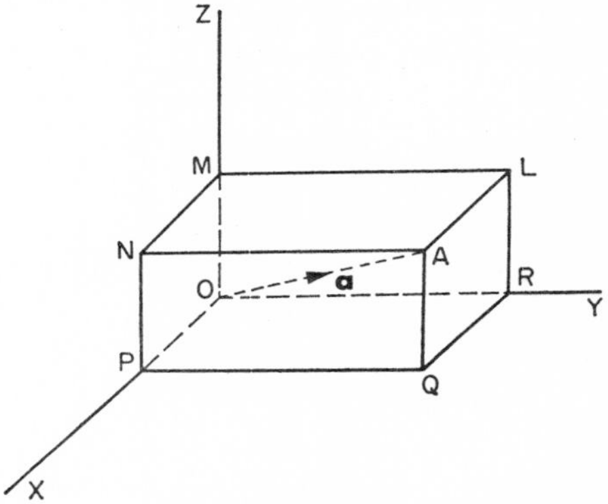

Fig. 9

$\mathbf{i}_1$, $\mathbf{i}_2$, $\mathbf{i}_3$, are unit vectors along *OX*, *OY*, *OZ* respectively and suppose $OP = a_1$, $OR = a_2$ and $OM = a_3$.

Since $$\overline{OA} = \overline{OP} + \overline{PQ} + \overline{QA}$$

i.e. $$\overline{OA} = \overline{OP} + \overline{OR} + \overline{OM},$$

then $$\mathbf{a} = a_1\mathbf{i}_1 + a_2\,\mathbf{i}_2 + a_3\,\mathbf{i}_3.$$

a_1, a_2, a_3 are known as the *components* of **a** in the directions *OX*, *OY*, *OZ* respectively.

It will in future be assumed that, when a fixed right-handed system of mutually perpendicular axes OX, OY, OZ, is being used, $\mathbf{i}_1$, $\mathbf{i}_2$, $\mathbf{i}_3$ will denote unit vectors in the directions OX, OY, OZ respectively, and that the components of any vector **a** or **b** . . . will be denoted by the corresponding suffixes, e.g. a_1, a_2, a_3; or b_1, b_2, b_3. . . .

§1.13. The Components of the Sum of Two Vectors

Let $\overline{OA}$, $\overline{AB}$ represent the vectors **a**, **b**, respectively where

$$\mathbf{a} = a_1\mathbf{i}_1 + a_2\mathbf{i}_2 + a_3\mathbf{i}_3$$

and

$$\mathbf{b} = b_1\mathbf{i}_1 + b_2\mathbf{i}_2 + b_3\mathbf{i}_3.$$

Complete the rectangular parallelepipeds $OA_1A_2A_3A_4A_5AA_6$ and $AB_1\ B_2\ B_3\ B_4\ B_5\ BB_6$.

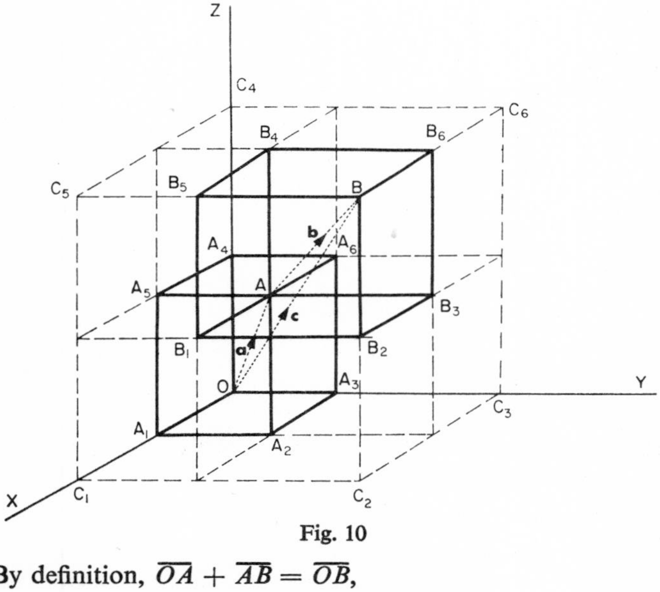

Fig. 10

By definition, $\overline{OA} + \overline{AB} = \overline{OB}$,

i.e.

$$\mathbf{a} + \mathbf{b} = \overline{OB}.$$

Complete the rectangular parallelepiped $OC_1C_2C_3C_4C_5BC_6$.

Then (see Fig. 10) the components of $\overline{OB}$ are $\overline{OC_1}$, $\overline{OC_3}$, $\overline{OC_4}$.

But
$$OC_1 = OA_1 + A_1C_1 = OA_1 + AB_1 = a_1 + b_1,$$
$$OC_3 = OA_3 + A_3C_3 = OA_3 + AB_3 = a_2 + b_2,$$
$$OC_4 = OA_4 + A_4C_4 = OA_4 + AB_4 = a_3 + b_3.$$

Hence,
$$\overline{OB} = (a_1 + b_1)\mathbf{i}_1 + (a_2 + b_2)\mathbf{i}_2 + (a_3 + b_3)\mathbf{i}_3,$$

i.e. the component of the sum of two vectors in any direction is equal to the sum of the components of the two vectors in the same direction.

The student would be well advised to draw his own diagram and to construct his own three-dimensional model to illustrate this result.

§1.14

By analogy with §1.13 above, if $\mathbf{a}, \mathbf{b}, \mathbf{c} \ldots$ is a system of vectors such that

$$\mathbf{a} = a_1\mathbf{i}_1 + a_2\mathbf{i}_2 + a_3\mathbf{i}_3$$
$$\mathbf{b} = b_1\mathbf{i}_1 + b_2\mathbf{i}_2 + b_3\mathbf{i}_3$$
$$\mathbf{c} = c_1\mathbf{i}_1 + c_2\mathbf{i}_2 + c_3\mathbf{i}_3$$
$$\ldots\ldots\ldots\ldots\ldots\ldots,$$

then
$$\mathbf{a} + \mathbf{b} + \mathbf{c} + \ldots = (a_1 + b_1 + c_1 + \ldots)\mathbf{i}_1 + (a_2 + b_2 + c_2 + \ldots)\mathbf{i}_2 + (a_3 + b_3 + c_3 + \ldots)\mathbf{i}_3.$$

§1.15. Modulus of a Vector in Terms of its Components in Three Mutually Perpendicular Directions

From Fig. 9,
$$OA^2 = OQ^2 + QA^2,$$
$$= OP^2 + PQ^2 + QA^2, \qquad (1)$$

i.e.
$$OA^2 = OP^2 + OR^2 + OM^2$$
$$= a_1^2 + a_2^2 + a_3^2$$

i.e. $$a = OA = \sqrt{(a_1^2 + a_2^2 + a_3^2)}.$$

Similarly from Fig. 10

$$OB = (a_1 + b_1)^2 + (a_2 + b_2)^2 + (a_3 + b_3)^2$$

i.e. $$|\mathbf{a} + \mathbf{b}| = \sqrt{\{(a_1 + b_1)^2 + (a_2 + b_2)^2 + (a_3 + b_3)^2},$$

and in general

$$|\mathbf{a} + \mathbf{b} + \mathbf{c} + \ldots| = \sqrt{\{(a_1 + b_1 + c_1 + \ldots)^2 + (a_2 + b_2 + c_2 + \ldots)^2 + (a_3 + b_3 + c_3 + \ldots)^2\}}$$

§1.16. Oblique Axes

The results of §§1.12, 1.13, 1.14 would be true whether the axes were rectangular or not, as the only properties used are those of any parallelepiped. The $\mathbf{i}_1$, $\mathbf{i}_2$, $\mathbf{i}_3$, notation is, however, reserved for use with rectangular axes only. The results of §1.15 are valid *only* when rectangular axes are used, since equation (1) of §1.15 depends upon the Theorem of Pythagoras.

§1.17. Direction Cosines and Direction Ratios

Suppose that the vector represented by $\overline{OA}$, makes angles α_1, α_2, α_3, with the rectangular axes OX, OY, OZ, respectively, and

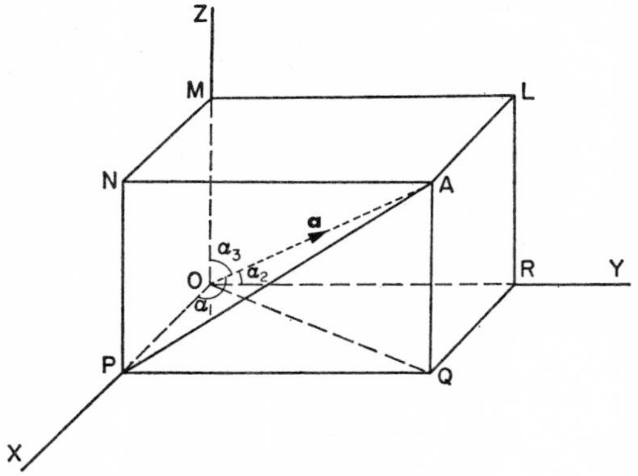

Fig. 11

suppose that a_1, a_2, a_3 are the components of **a**. Complete the rectangular parallelepiped *OPQRMNAL*.

Then in right-angled triangle *OPA*,

$$\left.\begin{aligned} &\cos \alpha_1 = \frac{OP}{OA} = \frac{a_1}{a}; \\ \text{Similarly,}\quad &\cos \alpha_2 = \frac{OR}{OA} = \frac{a_2}{a}, \\ \text{and}\quad &\cos \alpha_3 = \frac{OM}{OA} = \frac{a_3}{a}. \end{aligned}\right\} \qquad \text{(I)}$$

The three cosines are known as the *direction cosines* of the line *OA*, and are unique for a given line *OA*.

From equations (I),

$$\cos^2 \alpha_1 + \cos^2 \alpha_2 + \cos^2 \alpha_3 = \frac{a_1^2 + a_2^2 + a_3^2}{a^2} = \frac{a^2}{a^2} = 1.$$

Also, from equations (I),

$$\begin{aligned} \cos \alpha_1 : \cos \alpha_2 : \cos \alpha_3 &= a_1 : a_2 : a_3 \\ &= ka_1 : ka_2 : ka_3, \end{aligned}$$

where k is any convenient number. If $\lambda_1 = ka_1$, $\lambda_2 = ka_2$, $\lambda_3 = ka_3$ this result may be written

$$\cos \alpha_1 : \cos \alpha_2 : \cos \alpha_3 = \lambda_1 : \lambda_2 : \lambda_3.$$

A set of ratios of the form $\lambda_1 : \lambda_2 : \lambda_3$ is known as a set of *direction ratios* of the line *OA*. There is an infinite number of such sets of direction ratios, and for convenience a set is usually chosen so that λ_1, λ_2, λ_3 are as simple as possible to handle, e.g. integers, or simple surds.

Since $\qquad \cos \alpha_1 : \cos \alpha_2 : \cos \alpha_3 = \lambda_1 : \lambda_2 : \lambda_3,$

then $$\frac{\cos \alpha_1}{\lambda_1} = \frac{\cos \alpha_2}{\lambda_2} = \frac{\cos \alpha_3}{\lambda_3} = \frac{1}{\sqrt{(\lambda_1^2 + \lambda_2^2 + \lambda_3^2)}}.$$

Hence
$$\cos \alpha_1 = \frac{\lambda_1}{\sqrt{(\lambda_1^2 + \lambda_2^2 + \lambda_3^2)}},$$
$$\cos \alpha_2 = \frac{\lambda_2}{\sqrt{(\lambda_1^2 + \lambda_2^2 + \lambda_3^2)}},$$
$$\cos \alpha_3 = \frac{\lambda_3}{\sqrt{(\lambda_1^2 + \lambda_2^2 + \lambda_3^2)}}.$$

Thus, if a set of direction ratios of a line is known, the direction cosines can be calculated.

From §1.13 (see Fig. 10), the direction cosines of $(\mathbf{a} + \mathbf{b})$ are
$$\frac{a_1 + b_1}{|\mathbf{a} + \mathbf{b}|}, \quad \frac{a_2 + b_2}{|\mathbf{a} + \mathbf{b}|}, \quad \frac{a_3 + b_3}{|\mathbf{a} + \mathbf{b}|},$$
Similarly, the direction cosines of $(\mathbf{a} - \mathbf{b})$ are
$$\frac{a_1 - b_1}{|\mathbf{a} - \mathbf{b}|}, \quad \frac{a_2 - b_2}{|\mathbf{a} - \mathbf{b}|}, \quad \frac{a_3 - b_3}{|\mathbf{a} - \mathbf{b}|}.$$

§1.18. Geometrical Application

Suppose that **a**, **b** are the position vectors of two points A, B referred to a fixed origin O. Then
$$\overline{OA} = \mathbf{a} \quad \text{and} \quad \overline{OB} = \mathbf{b}.$$
Hence
$$\overline{AB} = \overline{AO} + \overline{OB}$$
$$= -\overline{OA} + \overline{OB}$$

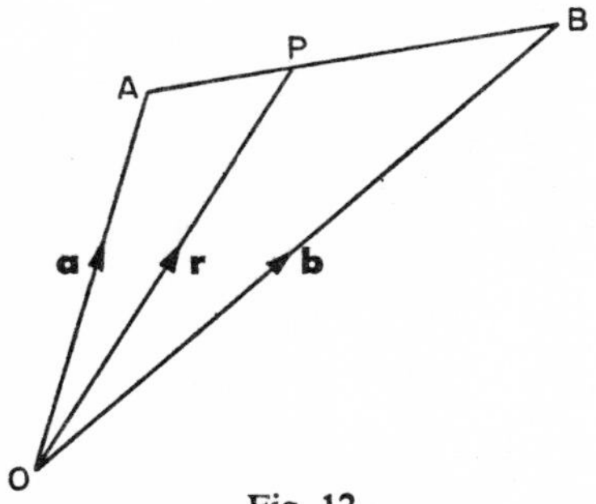

Fig. 12

i.e. $$\overline{AB} = \mathbf{b} - \mathbf{a}, \tag{I}$$

and $$AB = |\mathbf{b} - \mathbf{a}|$$

i.e. $$AB = \sqrt{\{(b_1 - a_1)^2 + (b_2 - a_2)^2 + (b_3 - a_3)^2\}} \tag{II}$$

The direction ratios of AB are $(b_1 - a_1) : (b_2 - a_2) : (b_3 - a_3)$. If P is any point in AB such that $AP : AB = t : 1$ and if $\mathbf{r}$ is the position vector of P, then

since $$AP : AB = t : 1,$$

therefore $$\overline{AP} = t\overline{AB},$$

i.e. $$\mathbf{r} - \mathbf{a} = t(\mathbf{b} - \mathbf{a}),$$

i.e. $$\mathbf{r} = \mathbf{a} + t(\mathbf{b} - \mathbf{a}). \tag{III}$$

which may be regarded as the vector equation of the line AB.

Further, suppose $AP : PB = \lambda : \mu$, then (see §1.10)

$$(\lambda + \mu)\mathbf{r} = \mu\mathbf{a} + \lambda\mathbf{b},$$

i.e. $$\mathbf{r} = \frac{\mu\mathbf{a} + \lambda\mathbf{b}}{\lambda + \mu}. \tag{IV}$$

§1.19. Centroid of a Triangle

Suppose $\mathbf{a}$, $\mathbf{b}$, $\mathbf{c}$ are the position vectors of the vertices A, B, C of a triangle referred to a fixed origin O. Let D be the mid-point of BC and let $\mathbf{d}$ be the position vector of D. Then from equation (IV) of §1.18,

$$\mathbf{d} = \frac{\mathbf{b} + \mathbf{c}}{2}.$$

If G, having position vector $\mathbf{g}$, is the centroid of triangle ABC, then G lies on AD and divides AD so that

$$AG : GD = 2 : 1$$

Hence $$\mathbf{g} = \frac{1(\mathbf{a}) + 2(\mathbf{d})}{2 + 1},$$

i.e. $$\mathbf{g} = \tfrac{1}{3}\{\mathbf{a} + \mathbf{b} + \mathbf{c}\}.$$

Example 3. *If* **a**, **b**, **c** *are the position vectors of the vertices A, B. C of parallelogram ABCD, find the position vector of D.*

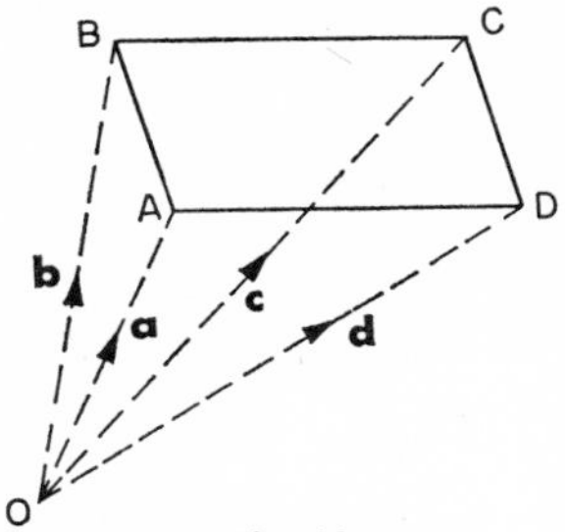

Fig. 13

Since AB, DC are equal and parallel,

$$\overline{DC} = \overline{AB}.$$

Then, if **d** is the position vector of D,

$$\mathbf{c} - \mathbf{d} = \mathbf{b} - \mathbf{a},$$

i.e.
$$\mathbf{d} = \mathbf{a} - \mathbf{b} + \mathbf{c}.$$

Example 4. *If the position vectors of the points A, B are* $2\mathbf{i}_1 + 4\mathbf{i}_2 - 5\mathbf{i}_3$, $3\mathbf{i}_1 + 2\mathbf{i}_2 + 7\mathbf{i}_3$, *find the magnitude and direction of* $\overline{AB}$.

Let **a** and **b** be the position vectors of A and B.

Then
$$\mathbf{a} = 2\mathbf{i}_1 + 4\mathbf{i}_2 - 5\mathbf{i}_3,$$

$$\mathbf{b} = 3\mathbf{i}_1 + 2\mathbf{i}_2 + 7\mathbf{i}_3.$$

Therefore
$$\overline{AB} = \mathbf{b} - \mathbf{a},$$

i.e.
$$\overline{AB} = \mathbf{i}_1 - 2\mathbf{i}_2 + 12\mathbf{i}_3.$$

Hence
$$AB^2 = 1 + 4 + 144.$$

Therefore
$$AB = \sqrt{(149)},$$

and the direction ratios of AB are $1 : (-2) : 12$.

Example 5. *Show that the points* $A(2, 6, 3)$, $B(1, 2, 7)$ *and* $C(3, 10, -1)$ *are collinear.*

Let $\mathbf{a}$, $\mathbf{b}$, $\mathbf{c}$ be the position vectors of A, B, C respectively.

Then
$$\mathbf{a} = 2\mathbf{i}_1 + 6\mathbf{i}_2 + 3\mathbf{i}_3,$$
$$\mathbf{b} = \mathbf{i}_1 + 2\mathbf{i}_2 + 7\mathbf{i}_3,$$
$$\mathbf{c} = 3\mathbf{i}_1 + 10\mathbf{i}_2 - \mathbf{i}_3.$$

Thus
$$\overline{AB} = \mathbf{b} - \mathbf{a} = -\mathbf{i}_1 - 4\mathbf{i}_2 + 4\mathbf{i}_3$$

and
$$\overline{AC} = \mathbf{c} - \mathbf{a} = \mathbf{i}_1 + 4\mathbf{i}_2 - 4\mathbf{i}_3.$$

Hence the direction ratios of AB, AC are both $1 : 4 : (-4)$; the two lines AB, AC have a common point A, and, therefore, the points A, B, C, must be collinear.

Example 6. *Show that the points whose position vectors are* $\mathbf{a} + 3\mathbf{b}$, $4\mathbf{a} - \mathbf{b}$, $7\mathbf{a} - 5\mathbf{b}$ *are collinear.*

Let P, Q, R be the points whose position vectors are $\mathbf{a} + 3\mathbf{b}$, $4\mathbf{a} - \mathbf{b}$, $7\mathbf{a} - 5\mathbf{b}$, respectively. Then
$$\overline{PQ} = 3\mathbf{a} - 4\mathbf{b},$$

and
$$\overline{PR} = 6\mathbf{a} - 8\mathbf{b};$$

i.e.
$$\overline{PR} = 2\overline{PQ}.$$

Hence $\overline{PQ}$, $\overline{PR}$ are two vectors both passing through the point P and having the same direction. Thus P, Q, R, are collinear.

Examples Ib

1. If the position vectors of two points P and Q are $7\mathbf{i}_1 + 3\mathbf{i}_2 - \mathbf{i}_3$ and $2\mathbf{i}_1 - 5\mathbf{i}_2 + 4\mathbf{i}_3$ respectively, find the magnitude and direction of the vector $\overline{PQ}$.

2. If the position vectors of the points A, B are $4\mathbf{i}_1 + 3\mathbf{i}_2 - 7\mathbf{i}_3$ and $12\mathbf{i}_1 + 6\mathbf{i}_2 - 2\mathbf{i}_3$ respectively, find the magnitude and direction of the vector $\overline{AB}$.

3. The position vectors of the points A, B, C, D are $\mathbf{a}$, $\mathbf{b}$, $2\mathbf{a} + 3\mathbf{b}$, $\mathbf{a} - 2\mathbf{b}$ respectively. Express the vectors $\overline{AC}$, $\overline{DB}$, $\overline{BC}$, $\overline{CA}$ in terms of $\mathbf{a}$, $\mathbf{b}$.

4. If $\mathbf{a}$, $\mathbf{b}$, $\mathbf{c}$ are the position vectors of the three vertices A, B, C of the regular hexagon $ABCDEF$, find the position vectors of the remaining three vertices in terms of $\mathbf{a}$, $\mathbf{b}$, $\mathbf{c}$.

5. Prove that if $ABCD$ is any quadrilateral (skew or plane) the lines joining the mid-points of opposite sides meet in a point, and that they bisect each other at that point.

6. Prove that if A, B, C are any three points, and G is another point such that

$$\overline{AG} + \overline{BG} + \overline{CG} = 0$$

then G is the point of intersection of the medians of the triangle ABC.

7. A, B, C, D are the vertices of a tetrahedron, and G is a point such that

$$\overline{AG} + \overline{BG} + \overline{CG} + \overline{DG} = 0.$$

Show that G lies on the line joining A to the centroid of triangle BCD. By showing similarly that G lies on the lines joining each of the other vertices to the centroid of the opposite face, show that if A', B', C', D' are the centroids of the faces BCD, CDA, ABD, ABC, respectively, the lines AA', BB', CC', DD' are concurrent and divide each other in the ratio 3 : 1.

8. Prove that the lines which join the mid-points of opposite edges of a tetrahedron are concurrent and bisect each other.

9. Show that the points whose position vectors are $\mathbf{a}$, $\mathbf{b}$, $3\mathbf{a} - 2\mathbf{b}$, are collinear.

10. Show that the points whose position vectors are $\mathbf{a}$, $\mathbf{b}$, $5\mathbf{a} - 4\mathbf{b}$, are collinear.

11. Show that the points whose position vectors are $\mathbf{a} + \mathbf{b}$, $2\mathbf{a} + 3\mathbf{b}$, $5\mathbf{a} + 9\mathbf{b}$, are collinear.

12. Show that the points (1, 3, 5), (2, −1, 3) and (4, −9, −1) are collinear.

13. Show that the points (2, −1, 3), (3, −5, 1), (−1, 11, 9) are collinear.

14. A, B are the points whose position vectors referred to the origin O are $\mathbf{i}_1 + 2\mathbf{i}_2 - \mathbf{i}_3$, $2\mathbf{i}_2 + 3\mathbf{i}_3$. C is the point which divides AB in the ratio 2 : 3; find the direction ratios and length of OC.

15. Show that if O, A, B, C are four points, and λ, μ, ν are three real numbers such that $\lambda + \mu + \nu = 0$, then if $\lambda\overline{OA} + \mu\overline{OB} + \nu\overline{OC} = 0$, the points A, B, C are collinear.

16. Prove that the four diagonals of a parallelepiped are concurrent and bisect each other. Prove also that the joins of the mid-points of opposite edges are concurrent at the same point and that they also bisect each other.

17. $ABCD$ is a parallelogram and P is the mid-point of BC. Prove that Q, the point of intersection of AP and BD is a point of trisection of both AP and BD.

18. A, B are the points whose position vectors are $\mathbf{i}_1 + 2\mathbf{i}_2 - 2\mathbf{i}_3$ and $2\mathbf{i}_1 + 3\mathbf{i}_2 + 6\mathbf{i}_3$. Forces of 6 lb wt and 21 lb wt. act along OA, OB. Write down the resultant force in the form $X\mathbf{i}_1 + Y\mathbf{i}_2 + Z\mathbf{i}_3$ and hence find the magnitude and direction of the resultant.

19. A, B, C are the points (0, 3, 4), (2, 3, 6), (2, 1, 2). Forces of 15 dyn, 14 dyn, 9 dyn act along OA, OB, OC respectively. Find the magnitude and direction of their resultant.

20. O is a vertex of a cube and forces of 2 lb wt., 4 lb wt., 3 lb wt. act along the diagonals OA, OB, OC of the three faces of the cube which contain O. Find the resultant of these forces.

21. $OABCDEFG$ is a cube in which OA, OC, OD are three mutually perpendicular edges, and DE is parallel to OA. Forces of 1 lb wt., 2 lb wt., 3 lb wt., 4 lb wt., act at O in the directions of the vectors $\overline{DA}$, $\overline{AC}$, $\overline{OF}$ and $\overline{DF}$ respectively. Find the magnitude and direction of the resultant of these forces.

CHAPTER II

§2.1

In the development of elementary algebra, as the concept of a number has been extended to include not only positive integers, but also fractions, directed numbers and complex numbers, so also have the operations of adding and multiplying two numbers been redefined at each stage so as to give to the operations meanings which are reasonable for application to the new members of the family of numbers, and yet not to conflict with previous definitions. The inverse processes of subtracting and dividing retain at each stage the property of being inverse processes, and do not, therefore, demand separate consideration. Thus, in elementary algebra, if d is the result of subtracting b from a, so that

$$d = a - b = a + (-b)$$

then

$$d + b = a$$

thus d is that number which when added to b gives the number a.

Similarly, if q is the result of dividing a by b, so that

$$q = \frac{a}{b}$$

then

$$qb = a$$

thus q is that number which when multiplied by b gives the number a.

§2.2

In defining the operations of addition and multiplication at each stage of elementary algebra, two laws have remained invariant, viz. the *commutative law* and the *associative law*. Thus for any two numbers a, b, real or complex,

$$a + b = b + a,$$

and

$$ab = ba,$$

also for three or more numbers, say a, b, c,

$$a + b + c = a + (b + c) = b + (c + a) = c + (a + b),$$

and

$$abc = a(bc) = b(ca) = c(ab).$$

The operation of multiplication obeys also the *distributive law* so that

$$a(b + c) = ab + ac.$$

In the operations of subtraction and division neither the commutative nor the associative laws are obeyed but the operations of subtraction and division are completely defined, as stated in §2.1, by their characteristics of being inverse operations, and hence if the operations of addition and multiplication can be performed, the results of subtraction and division can always be found.

§2.3

It is convenient at this stage to examine the extent to which the

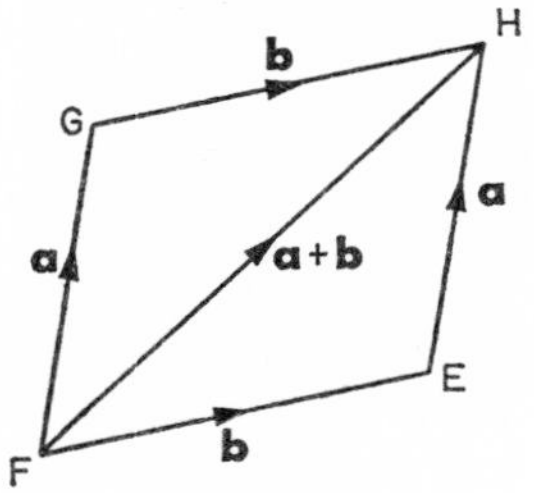

Fig. 14

definition of addition adopted in vector algebra conforms with the definition of addition adopted in real and complex algebra.

If in Fig. 2 of §1.7, the parallelogram *EFGH* is completed (see Fig. 14), then by definition,

$$\overline{FG} + \overline{GH} = \overline{FH} = \overline{FE} + \overline{EH}$$

i.e.

$$\mathbf{a} + \mathbf{b} = \overline{FH} = \mathbf{b} + \mathbf{a}.$$

Thus the sum of **a** and **b** obeys the commutative law (see also §1.6). Consider now the sum of three vectors **a, b, c.** Let $\overline{OA}$, $\overline{AB}$, $\overline{BC}$ represent **a, b, c** respectively (see Fig. 15).

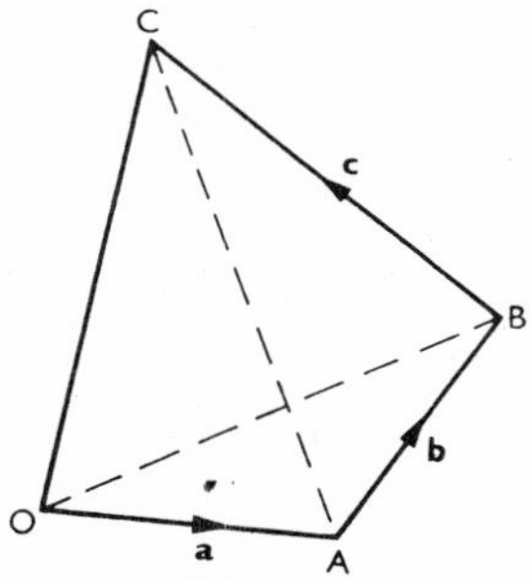

Fig. 15

Then

$$\mathbf{a} + \mathbf{b} = \overline{OA} + \overline{AB} = \overline{OB},$$

and

$$(\mathbf{a} + \mathbf{b}) + \mathbf{c} = \overline{OB} + \overline{BC} = \overline{OC}.$$

Also

$$\mathbf{b} + \mathbf{c} = \overline{AB} + \overline{BC} = \overline{AC},$$

and

$$\mathbf{a} + (\mathbf{b} + \mathbf{c}) = \overline{OA} + \overline{AC} = \overline{OC}.$$

Thus

$$(\mathbf{a} + \mathbf{b}) + \mathbf{c} = \mathbf{a} + (\mathbf{b} + \mathbf{c}).$$

Since the addition of two vectors is commutative,

$$\mathbf{a} + (\mathbf{b} + \mathbf{c}) = (\mathbf{b} + \mathbf{c}) + \mathbf{a} = \mathbf{b} + (\mathbf{c} + \mathbf{a}) = (\mathbf{c} + \mathbf{a}) + \mathbf{b}.$$

Hence

$$\mathbf{a} + (\mathbf{b} + \mathbf{c}) = (\mathbf{a} + \mathbf{b}) + \mathbf{c} = (\mathbf{c} + \mathbf{a}) + \mathbf{b},$$

i.e. the associative law holds for the addition of three vectors. This argument can be extended to the addition of four or more vectors.

§2.4

It remains to discuss whether the further development of vector algebra requires an operation in any way analogous to the operation of multiplication as used in real and complex algebra; and if such an operation is required, how to define it. An examination of elementary mechanics provides two examples in which vector quantities are combined in such a way that the result is proportional to the arithmetical product of their moduli, but as the result in the one case is scalar, and in the other a vector, the idea of two forms of product, a scalar product and a vector product, is suggested.

§2.5. Scalar Product of Two Vectors

Consider the work done by a force **F** as its point of application is displaced from the point A to the point B. If $\overline{AB} = \mathbf{s}$, the work

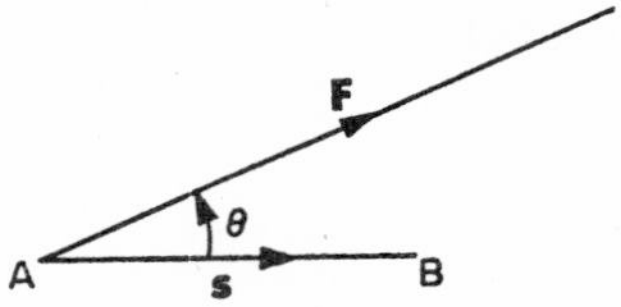

Fig. 16

done is the scalar quantity $(s \cos \theta)F$ where θ is the angle made by the direction of **F** with the direction of **s**. This suggests that the scalar result of multiplying **a** by **b** should be $ab \cos \theta$ where θ is

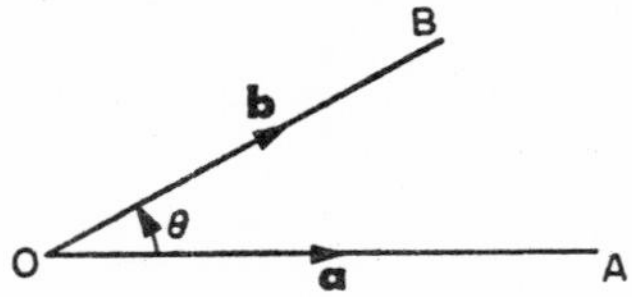

Fig. 17

the angle made by the direction of **b** with the direction of **a**. If this definition is adopted, then the scalar result of multiplying **b** by **a** should be $ba \cos(-\theta)$. But $ba = ab$ and $\cos(-\theta) = \cos \theta$. Hence the scalar result of multiplying **a** by **b** is the same as the scalar result of multiplying **b** by **a**. This form of the product of **a** and **b** is known as the *scalar product* of **a** and **b** and is written **a** . **b**, and read as "**a** dot **b**" or "the scalar product of **a** and **b**". It has been shown that $\mathbf{a} \cdot \mathbf{b} = \mathbf{b} \cdot \mathbf{a}$.

Hence

$$\mathbf{a} \cdot \mathbf{b} = ab \cos \theta = \mathbf{b} \cdot \mathbf{a}.$$

Thus the scalar product of two vectors obeys the commutative law.

Adopting this definition of a scalar product, the work done by the force **F** when the displacement of its point of application is **s** is $\mathbf{s} \cdot \mathbf{F}$.

§2.6. Vector Product of Two Vectors

Suppose A is the point whose position vector with respect to an origin O is **s**, and suppose **F** is a force acting through A along the line AB. Then the points O, A, B define a plane. Let θ be the

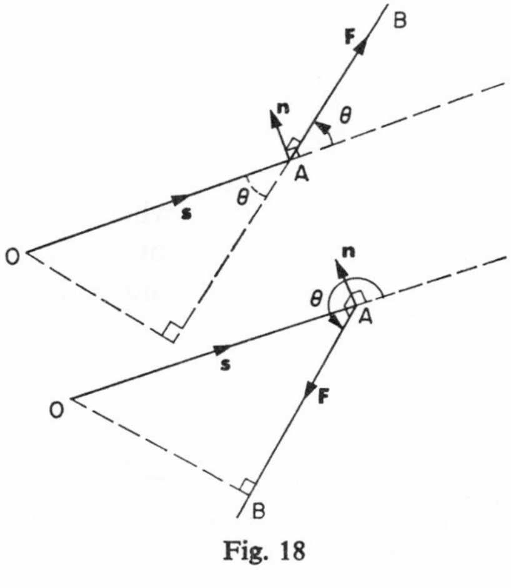

Fig. 18

angle which $\overline{AB}$ makes with $\overline{OA}$ and let **n** be the unit vector normal to the plane of O,A,B such that **s**, **F**, **n** form a right-handed system of vectors (see §1.12). The moment of **F** about O is of magnitude $(s \sin \theta)F$, i.e. $sF \sin \theta$; this moment is considered to be positive if its turning effect is in the direction in which θ increases, i.e. if $0 < \theta < \pi$; the moment of **F** about O is considered to be negative if its turning effect is in the direction in which θ decreases, i.e. if $\pi < \theta < 2\pi$. Thus the vector $(sF \sin \theta)\mathbf{n}$ has the magnitude of the moment of **F** about O; moreover the vector $(sF \sin \theta)\mathbf{n}$ is positive for $0 < \theta < \pi$ and negative for $\pi < \theta < 2\pi$, thus it has the *sign* of the moment of **F** about O. It is, therefore, convenient to say that the moment of **F** about O is the vector $(sF \sin \theta)\mathbf{n}$, where a vector in one line implies a turning effect in a plane perpendicular to that line, in a sense determined by the right-hand screw rule. This suggests that the vector result of multiplying the vector **a** by the vector **b** should be the vector $(ab \sin \theta)\mathbf{n}$,where **a**, **b**, **n** form a right-handed system of vectors and θ is the angle made with the direction of **a** by the direction of **b**. This form of the product **a** and **b** is known as the vector product of **a** and **b**, and is written $\mathbf{a} \times \mathbf{b}$. The expression $\mathbf{a} \times \mathbf{b}$ is read as "**a** cross **b**" or "the vector product of **a** and **b**". If this definition of a vector product is adopted, the vector product $\mathbf{b} \times \mathbf{a}$ should be equal to $\{ba \sin(-\theta)\}\mathbf{n}$, i.e. $(-ab \sin \theta)\mathbf{n}$, i.e. $-\mathbf{a} \times \mathbf{b}$,

i.e. $$\mathbf{a} \times \mathbf{b} = (ab \sin \theta)\mathbf{n} = -\mathbf{b} \times \mathbf{a}.$$

Thus the vector product of two vectors does *not* obey the commutative law.

Adopting this definition of a vector product the moment of the force F about the origin is $\mathbf{s} \times \mathbf{F}$ where **s** is the position vector of the point of application of **F**.

§2.7. Distributive Law—Scalar Product

(i) By definition, if θ is the angle between the vectors **a** and **b**,

$$\begin{aligned} \mathbf{a} \cdot \mathbf{b} &= ab \cos \theta \\ &= a(b \cos \theta) \end{aligned}$$

Similarly, if ϕ is the angle between the vectors **a** and **c**,

$$\mathbf{a} \cdot \mathbf{c} = ac \cos \phi$$
$$= a(c \cos \phi)$$

therefore
$$\mathbf{a} \cdot \mathbf{b} + \mathbf{a} \cdot \mathbf{c} = a(b \cos \theta) + a(c \cos \phi)$$
$$= a(b \cos \theta + c \cos \phi)$$

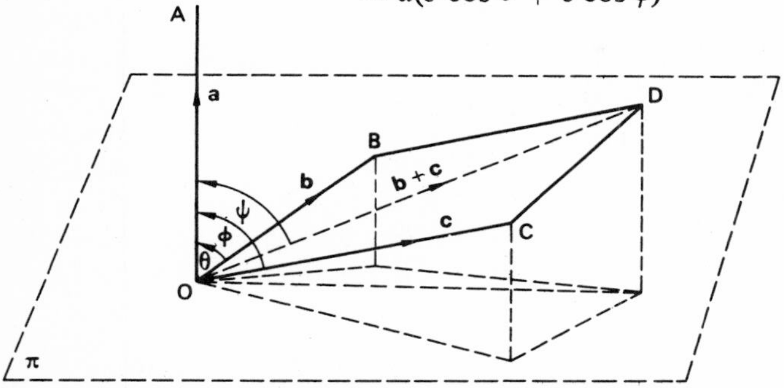

Fig. 19

If **r** is the sum of **b** and **c** making an angle ψ with **a**, then

$$\mathbf{a} \cdot \mathbf{r} = ar \cos \psi$$
$$= a(r \cos \psi)$$

But
$$\mathbf{r} = \mathbf{b} + \mathbf{c}.$$

hence, by §1.13, resolving in the direction of **a**,

$$r \cos \psi = b \cos \theta + c \cos \phi$$

therefore
$$\mathbf{a} \cdot \mathbf{r} = a(r \cos \psi)$$
$$= a(b \cos \theta + c \cos \phi)$$
$$= \mathbf{a} \cdot \mathbf{b} + \mathbf{a} \cdot \mathbf{c},$$

i.e.
$$\mathbf{a} \cdot (\mathbf{b} + \mathbf{c}) = \mathbf{a} \cdot \mathbf{b} + \mathbf{a} \cdot \mathbf{c}.$$

This argument can be extended to include any number of vectors, so that

$$\mathbf{a} \cdot (\mathbf{b} + \mathbf{c} + \mathbf{d} + \ldots) = \mathbf{a} \cdot \mathbf{b} + \mathbf{a} \cdot \mathbf{c} + \mathbf{a} \cdot \mathbf{d} + \ldots$$

Thus the distributive law is obeyed in the scalar multiplication of vectors.

(ii) The result of §2.7 (i) can be extended to include a product of the form

$$(\mathbf{a} + \mathbf{b} + \mathbf{c} + \ldots) \cdot (\mathbf{p} + \mathbf{q} + \mathbf{r} + \ldots).$$

For suppose that $\mathbf{a} + \mathbf{b} + \mathbf{c} + \ldots = \mathbf{A}$, then

$$\begin{aligned}(\mathbf{a} + \mathbf{b} + \mathbf{c} + \ldots) &\cdot (\mathbf{p} + \mathbf{q} + \mathbf{r} + \ldots)\\ &= \mathbf{A} \cdot (\mathbf{p} + \mathbf{q} + \mathbf{r} + \ldots)\\ &= \mathbf{A} \cdot \mathbf{p} + \mathbf{A} \cdot \mathbf{q} + \mathbf{A} \cdot \mathbf{r} + \ldots\end{aligned}$$

But

$$\begin{aligned}\mathbf{A} \cdot \mathbf{p} &= \mathbf{p} \cdot \mathbf{A}\\ &= \mathbf{p} \cdot (\mathbf{a} + \mathbf{b} + \mathbf{c} + \ldots)\\ &= \mathbf{p} \cdot \mathbf{a} + \mathbf{p} \cdot \mathbf{b} + \mathbf{p} \cdot \mathbf{c} + \ldots\\ &= \mathbf{a} \cdot \mathbf{p} + \mathbf{b} \cdot \mathbf{p} + \mathbf{c} \cdot \mathbf{p} + \ldots\end{aligned}$$

Similarly

$$\mathbf{A} \cdot \mathbf{q} = \mathbf{a} \cdot \mathbf{q} + \mathbf{b} \cdot \mathbf{q} + \mathbf{c} \cdot \mathbf{q} + \ldots, \text{ etc.}$$

Hence

$$\begin{aligned}(\mathbf{a} + \mathbf{b} + \mathbf{c} + \ldots) &\cdot (\mathbf{p} + \mathbf{q} + \mathbf{r} + \ldots)\\ &= \mathbf{a} \cdot \mathbf{p} + \mathbf{b} \cdot \mathbf{p} + \mathbf{c} \cdot \mathbf{p} + \ldots\\ &\quad + \mathbf{a} \cdot \mathbf{q} + \mathbf{b} \cdot \mathbf{q} + \mathbf{c} \cdot \mathbf{q} + \ldots\\ &\quad + \mathbf{a} \cdot \mathbf{r} + \ldots\\ &\quad \text{etc.}\end{aligned}$$

§2.8

By definition, if the vector **b** makes an angle θ with the vector **a**, and n is a real number

$$n(\mathbf{a} \cdot \mathbf{b}) = nab \cos \theta$$

also

$$(n\mathbf{a}) \cdot \mathbf{b} = nab \cos \theta$$

and

$$\mathbf{a} \cdot (n\mathbf{b}) = nab \cos \theta.$$

Hence

$$n(\mathbf{a} \cdot \mathbf{b}) = (n\mathbf{a}) \cdot \mathbf{b} = \mathbf{a} \cdot (n\mathbf{b}) = nab \cos \theta.$$

§2.9. Distributive Law—Vector Product

(i) Suppose the vectors **a** and **b** are represented by $\overline{OA}$ and $\overline{OB}$ and suppose Π is the plane through O perpendicular to OA. Let BB' be the perpendicular from B to Π and let **b**′ be the vector represented by $\overline{OB'}$. Then if **b** makes an angle θ with **a**,

$$b' = b \sin \theta$$

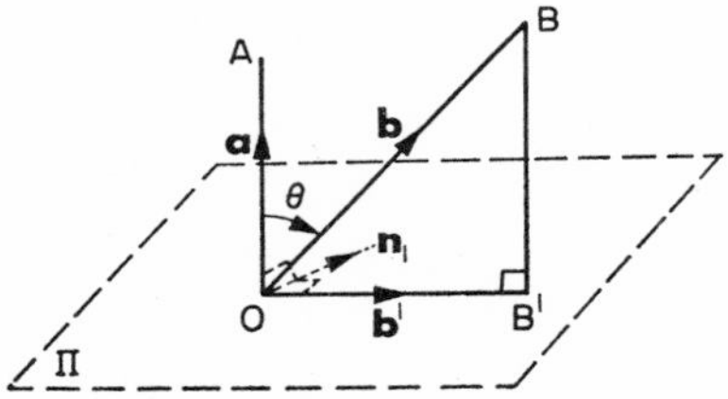

Fig. 20

Suppose $\mathbf{n}_1$ is the unit vector perpendicular to **a**, **b** so that **a**, **b**, $\mathbf{n}_1$ form a right-handed system, then

$$\mathbf{a} \times \mathbf{b} = (ab \sin \theta)\mathbf{n}_1.$$

Also

$$\mathbf{a} \times \mathbf{b}' = \left(ab' \sin \frac{\pi}{2}\right)\mathbf{n}_1$$

$$= (ab')\mathbf{n}_1$$

$$= (ab \sin \theta)\mathbf{n}_1.$$

i.e.

$$\mathbf{a} \times \mathbf{b} = \mathbf{a} \times \mathbf{b}' = ab'\mathbf{n}_1. \quad (1)$$

Suppose **c** is a third vector represented by $\overline{OC}$. Let CC' be drawn perpendicular to Π and let **c**′ be the vector represented by $\overline{OC'}$. Complete the parallelogram $OBDC$ and draw DD' perpendicular to Π (see Fig. 21). Since OC, BD are equal, and equally inclined to Π, then OC' and $B'D'$ are equal. Similarly OB' and $C'D'$ are equal. Hence $OB'D'C'$ is a parallelogram. Let **c**′, **d**, **d**′ be the vectors represented by $\overline{OC'}$, $\overline{OD}$, $\overline{OD'}$ respectively. Since $OBDC$, $OB'D'C'$ are parallelograms,

$$\mathbf{d} = \mathbf{b} + \mathbf{c} \quad \text{and} \quad \mathbf{d}' = \mathbf{b}' + \mathbf{c}'$$

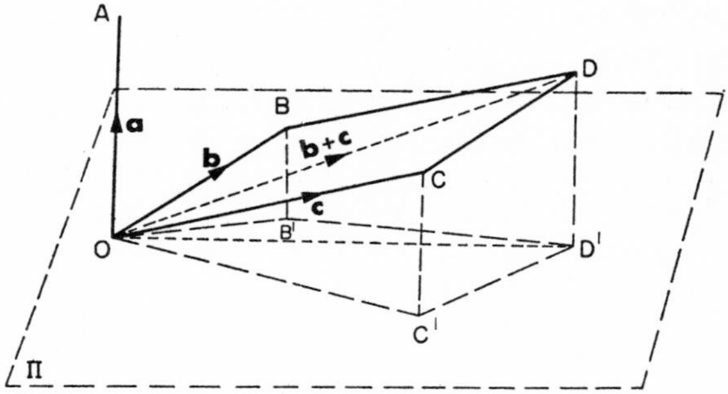

Fig. 21

Using the result (1) above,

$$\mathbf{a} \times \mathbf{b} = \mathbf{a} \times \mathbf{b}' = ab'\mathbf{n}_1,$$

$$\mathbf{a} \times \mathbf{c} = \mathbf{a} \times \mathbf{c}' = ac'\mathbf{n}_2,$$

and $$\mathbf{a} \times \mathbf{d} = \mathbf{a} \times \mathbf{d}' = ad'\mathbf{n}_3,$$

where $\mathbf{n}_1$, $\mathbf{n}_2$, $\mathbf{n}_3$ are three unit vectors in plane Π perpendicular to OB', OC', OD' respectively. Lines OB'', OC'', OD'' can be drawn in plane Π perpendicular to OB', OC', OD' respectively, so that $\overline{OB}'' = ab'\mathbf{n}_1$, $\overline{OC}'' = ac'\mathbf{n}_2$, and $\overline{OD}'' = ad'\mathbf{n}_3$. Since the sides of the quadrilateral $OB''D''C''$ are proportional to the sides of parallelogram $OB'D'C'$, and the angles of quadrilateral $OB''D''C''$ are equal to the angles of parallelogram $OB'D'C'$, then the quadrilateral $OB''D''C''$ is also a parallelogram, and

$$\overline{OD}'' = \overline{OB}'' + \overline{OC}'',$$

i.e. $$ad'\mathbf{n}_3 = ab'\mathbf{n}_1 + ac'\mathbf{n}_2$$

i.e. $$\mathbf{a} \times \mathbf{d} = \mathbf{a} \times \mathbf{b} + \mathbf{a} \times \mathbf{c},$$

i.e. $$\mathbf{a} \times (\mathbf{b} + \mathbf{c}) = \mathbf{a} \times \mathbf{b} + \mathbf{a} \times \mathbf{c}.$$

As in the case of the scalar product, this argument can be extended to include any number of vectors so that

$$\mathbf{a} \times (\mathbf{b} + \mathbf{c} + \mathbf{e} + \ldots) = \mathbf{a} \times \mathbf{b} + \mathbf{a} \times \mathbf{c} + \mathbf{a} \times \mathbf{e} + \ldots$$

Hence the distributive law is obeyed in the vector multiplication of vectors.

(ii) The result of §2.9 (i) can be extended to include a product of the form

$$(\mathbf{a} + \mathbf{b} + \mathbf{c} + \ldots) \times (\mathbf{p} + \mathbf{q} + \mathbf{r} + \ldots).$$

For suppose that $\mathbf{a} + \mathbf{b} + \mathbf{c} + \ldots = \mathbf{A}$, then

$$\begin{aligned}(\mathbf{a} + \mathbf{b} + \mathbf{c} + \ldots) &\times (\mathbf{p} + \mathbf{q} + \mathbf{r} + \ldots)\\ &= \mathbf{A} \times (\mathbf{p} + \mathbf{q} + \mathbf{r} + \ldots)\\ &= \mathbf{A} \times \mathbf{p} + \mathbf{A} \times \mathbf{q} + \mathbf{A} \times \mathbf{r} + \ldots.\end{aligned}$$

Now
$$\begin{aligned}\mathbf{A} \times \mathbf{p} &= -\mathbf{p} \times \mathbf{A}\\ &= -\mathbf{p} \times (\mathbf{a} + \mathbf{b} + \mathbf{c} + \ldots)\\ &= -\mathbf{p} \times \mathbf{a} - \mathbf{p} \times \mathbf{b} - \mathbf{p} \times \mathbf{c} - \ldots\\ &= \mathbf{a} \times \mathbf{p} + \mathbf{b} \times \mathbf{p} + \mathbf{c} \times \mathbf{p} + \ldots.\end{aligned}$$

Similarly,

$$\mathbf{A} \times \mathbf{q} = \mathbf{a} \times \mathbf{q} + \mathbf{b} \times \mathbf{q} + \mathbf{c} \times \mathbf{q} + \ldots \text{ etc.}$$

Hence

$$\begin{aligned}(\mathbf{a} + \mathbf{b} + \mathbf{c} + \ldots) &\times (\mathbf{p} + \mathbf{q} + \mathbf{r} + \ldots)\\ &= \mathbf{a} \times \mathbf{p} + \mathbf{b} \times \mathbf{p} + \mathbf{c} \times \mathbf{p} + \ldots\\ &\quad + \mathbf{a} \times \mathbf{q} + \mathbf{b} \times \mathbf{q} + \mathbf{c} \times \mathbf{q} + \ldots\\ &\quad\quad + \mathbf{a} \times \mathbf{r} + \mathbf{b} \times \mathbf{r} + \ldots\\ &\quad\quad\quad + \ldots\end{aligned}$$

§2.10

If the vector **b** makes an angle θ with the vector **a**, and m is a real number, and **n** is the unit vector perpendicular to **a** and **b** so that **a**, **b**, **n** form a right-handed system, then, by definition,

$$m(\mathbf{a} \times \mathbf{b}) = (mab \sin \theta)\mathbf{n}.$$

Also
$$(m\mathbf{a}) \times \mathbf{b} = (mab \sin \theta)\mathbf{n},$$

and
$$\mathbf{a} \times (m\mathbf{b}) = (mab \sin \theta)\mathbf{n}.$$

Hence
$$m(\mathbf{a} \times \mathbf{b}) = m\mathbf{a} \times \mathbf{b} = \mathbf{a} \times m\mathbf{b} = (mab \sin \theta)\mathbf{n}.$$

§2.11

With the usual notation, if $\mathbf{i}_1$, $\mathbf{i}_2$, $\mathbf{i}_3$ are unit vectors in the directions of the right-handed system of mutually perpendicular axes Ox, Oy, Oz, it follows from the definitions of §§2.5 and 2.6 that

$$\mathbf{i}_1 \,.\, \mathbf{i}_1 = \mathbf{i}_2 \,.\, \mathbf{i}_2 = \mathbf{i}_3 \,.\, \mathbf{i}_3 = 1 \qquad \text{(I)}$$

and $$\mathbf{i}_1 \,.\, \mathbf{i}_2 = \mathbf{i}_2 \,.\, \mathbf{i}_1 = \mathbf{i}_2 \,.\, \mathbf{i}_3 = \mathbf{i}_3 \,.\, \mathbf{i}_2 = \mathbf{i}_3 \,.\, \mathbf{i}_1 = \mathbf{i}_1 \,.\, \mathbf{i}_3 = 0 \qquad \text{(II)}$$

Also, $$\mathbf{i}_1 \times \mathbf{i}_1 = \mathbf{i}_2 \times \mathbf{i}_2 = \mathbf{i}_3 \times \mathbf{i}_3 = 0 \qquad \text{(III)}$$

and $$\mathbf{i}_1 \times \mathbf{i}_2 = -\mathbf{i}_2 \times \mathbf{i}_1 = \mathbf{i}_3 \qquad \text{(IV)}$$

$$\mathbf{i}_2 \times \mathbf{i}_3 = -\mathbf{i}_3 \times \mathbf{i}_2 = \mathbf{i}_1 \qquad \text{(V)}$$

$$\mathbf{i}_3 \times \mathbf{i}_1 = -\mathbf{i}_1 \times \mathbf{i}_3 = \mathbf{i}_2 \qquad \text{(VI)}$$

§2.12

It follows from (I) and (II) of §2.11 that if the vectors **a**, **b** are expressed in the forms

$$\mathbf{a} = a_1\mathbf{i}_1 + a_2\mathbf{i}_2 + a_3\mathbf{i}_3,$$

and $$\mathbf{b} = b_1\mathbf{i}_1 + b_2\mathbf{i}_2 + b_3\mathbf{i}_3,$$

then $$\mathbf{a} \,.\, \mathbf{b} = (a_1\mathbf{i}_1 + a_2\mathbf{i}_2 + a_3\mathbf{i}_3) \,.\, (b_1\mathbf{i}_1 + b_2\mathbf{i}_2 + b_3\mathbf{i}_3)$$

i.e. $$\mathbf{a} \,.\, \mathbf{b} = a_1b_1 + a_2b_2 + a_3b_3. \qquad \text{(VII)}$$

Also, if **b** makes an angle θ with **a**,

$$\mathbf{a} \,.\, \mathbf{b} = ab \cos \theta$$

therefore $$ab \cos \theta = a_1b_1 + a_2b_2 + a_3b_3,$$

i.e. $$\cos \theta = \frac{a_1b_1 + a_2b_2 + a_3b_3}{ab}.$$

But $a = (a_1^2 + a_2^2 + a_3^2)^{\frac{1}{2}}$ and $b = (b_1^2 + b_2^2 + b_3^2)^{\frac{1}{2}}$ and the direction ratios of the lines representing **a** and **b** are $a_1 : a_2 : a_3$ and $b_1 : b_2 : b_3$ respectively. Hence the angle between two lines whose direction ratios are $a_1 : a_2 : a_3$ and $b_1 : b_2 : b_3$ is θ where

$$\theta = \cos^{-1} \frac{a_1b_1 + a_2b_2 + a_3b_3}{(a_1^2 + a_2^2 + a_3^2)^{\frac{1}{2}}(b_1^2 + b_2^2 + b_3^2)^{\frac{1}{2}}}. \qquad \text{(VIII)}$$

§2.13

Using the notation of §2.12 and the results III, IV, V, VI of §2.11, it follows that

$$\mathbf{a} \times \mathbf{b} = (a_1\mathbf{i}_1 + a_2\mathbf{i}_2 + a_3\mathbf{i}_3) \times (b_1\mathbf{i}_1 + b_2\mathbf{i}_2 + b_3\mathbf{i}_3)$$

i.e. $$\mathbf{a} \times \mathbf{b} = (a_2b_3 - a_3b_2)\mathbf{i}_1 + (a_3b_1 - a_1b_3)\mathbf{i}_2 + (a_1b_2 - a_2b_1)\mathbf{i}_3 \qquad \text{(IX)}$$

From IX it can be seen that the components of the vector $\mathbf{a} \times \mathbf{b}$ are the second order determinants contained in the matrix

$$\begin{pmatrix} a_1 & a_2 & a_3 \\ b_1 & b_2 & b_3 \end{pmatrix}.$$

Alternatively the result IX could be obtained by expanding the determinant

$$\begin{vmatrix} \mathbf{i}_1 & \mathbf{i}_2 & \mathbf{i}_3 \\ a_1 & a_2 & a_3 \\ b_1 & b_2 & b_3 \end{vmatrix} \qquad \text{(X)}$$

treating $\mathbf{i}_1$, $\mathbf{i}_2$, $\mathbf{i}_3$ as if they were real algebraic quantities. This is a quick way of obtaining the result, and an easy symmetrical form for remembering it, but the student should bear in mind the fact that $\mathbf{i}_1$, $\mathbf{i}_2$, $\mathbf{i}_3$ are *not* real algebraic quantities, and that the determinant X is no more than a convenient symbolic form of the result expressed in IX.

Example 1. *Find the scalar product of the two vectors* $2\mathbf{i}_1 + 3\mathbf{i}_2 - 4\mathbf{i}_3$ and $4\mathbf{i}_1 - 2\mathbf{i}_2 - 3\mathbf{i}_3$.

Let $$\mathbf{a} = 2\mathbf{i}_1 + 3\mathbf{i}_2 - 4\mathbf{i}_3,$$

and $$\mathbf{b} = 4\mathbf{i}_1 - 2\mathbf{i}_2 - 3\mathbf{i}_3.$$

Then $$\mathbf{a} \,.\, \mathbf{b} = (2)(4) + (3)(-2) + (-4)(-3),$$
$$= 8 - 6 + 12$$

i.e. $$\mathbf{a} \,.\, \mathbf{b} = 14.$$

Example 2. *Find the vector product of the two vectors*

$$2\mathbf{i}_1 + \mathbf{i}_2 - 5\mathbf{i}_3 \textit{ and } 3\mathbf{i}_1 - 4\mathbf{i}_2 + 6\mathbf{i}_3.$$

Let

$$\mathbf{a} = 2\mathbf{i}_1 + \mathbf{i}_2 - 5\mathbf{i}_3$$

$$\mathbf{b} = 3\mathbf{i}_1 - 4\mathbf{i}_2 + 6\mathbf{i}_3$$

Then

$$\mathbf{a} \times \mathbf{b} = (6-20)\mathbf{i}_1 + (-15-12)\mathbf{i}_2 + (-8-3)\mathbf{i}_3 \qquad \begin{vmatrix} \mathbf{i}_1 & \mathbf{i}_2 & \mathbf{i}_3 \\ 2 & 1 & -5 \\ 3 & -4 & 6 \end{vmatrix}$$

i.e. $\mathbf{a} \times \mathbf{b} = -14\mathbf{i}_1 - 27\mathbf{i}_2 - 11\mathbf{i}_3$

Example 3. *Find the work done by a force of* 3 *lb wt. acting parallel to the line AB, if its point of application moves from A to C, where A, B, C are the points whose position vectors are* $\mathbf{i}_1 + 4\mathbf{i}_2 + 3\mathbf{i}_3$, $3\mathbf{i}_1 + 2\mathbf{i}_2 + 6\mathbf{i}_3$, $5\mathbf{i}_1 + 7\mathbf{i}_2 + \mathbf{i}_3$ *respectively, the distances being measured in feet.*

Let **a**, **b**, **c** be the position vectors of *A*, *B*, *C* where

$$\mathbf{a} = \mathbf{i}_1 + 4\mathbf{i}_2 + 3\mathbf{i}_3$$

$$\mathbf{b} = 3\mathbf{i}_1 + 2\mathbf{i}_2 + 6\mathbf{i}_3$$

$$\mathbf{c} = 5\mathbf{i}_1 + 7\mathbf{i}_2 + \mathbf{i}_3$$

Then

$$\overline{AB} = \mathbf{b} - \mathbf{a}$$

i.e.

$$\overline{AB} = 2\mathbf{i}_1 - 2\mathbf{i}_2 + 3\mathbf{i}_3$$

Hence the direction cosines of $\overline{AB}$, the line of action of the force of 3 lb wt., are $2/\sqrt{17}$, $-2\sqrt{17}$, $3/\sqrt{17}$, and if **F** denotes the force,

$$\mathbf{F} = 3(2\mathbf{i}_1 - 2\mathbf{i}_2 + 3\mathbf{i}_3)/\sqrt{17}$$

The displacement of the point of application of **F** is $\overline{AC}$ where

$$\overline{AC} = \mathbf{c} - \mathbf{a}$$

$$= 4\mathbf{i}_1 + 3\mathbf{i}_2 - 2\mathbf{i}_3,$$

Then the work done by **F** is W where

$$W = \overline{AC} \,.\, \mathbf{F}$$

$$= (4\mathbf{i}_1 + 3\mathbf{i}_2 - 2\mathbf{i}_3) \,.\, \frac{3}{\sqrt{17}} (2\mathbf{i}_1 - 2\mathbf{i}_2 + 3\mathbf{i}_3) \text{ft lb wt.}$$

$$= \frac{3}{\sqrt{17}} (8 - 6 - 6) \text{ft lb wt.}$$

i.e. $W = -12/\sqrt{17}$ ft lb wt.

Example 4. *Find the moment about a point C of a force of 7 lb wt. acting along the line AB, where A, B, C, are the points* (1, 2, −1), (3, 5, 4), (2, 3, 6) *respectively, the distances being measured in feet.*

Let **a**, **b**, **c**, be the position vectors of the points A, B, C, then

$$\mathbf{a} = \mathbf{i}_1 + 2\mathbf{i}_2 - \mathbf{i}_3$$

$$\mathbf{b} = 3\mathbf{i}_1 + 5\mathbf{i}_2 + 4\mathbf{i}_3$$

$$\mathbf{c} = 2\mathbf{i}_1 + 3\mathbf{i}_2 + 6\mathbf{i}_3$$

Let **F** be the force of 7 lb wt. acting along AB.

Since $\overline{AB} = \mathbf{b} - \mathbf{a}$

$$= 2\mathbf{i}_1 + 3\mathbf{i}_2 + 5\mathbf{i}_3$$

therefore $\mathbf{F} = \frac{7}{\sqrt{38}} (2\mathbf{i}_1 + 3\mathbf{i}_2 + 5\mathbf{i}_3)$

Since any point in the line of action of a force may be taken to be its point of application, suppose A is taken to be the point of application of **F**. Then the moment of **F** about C is $\overline{CA} \times \mathbf{F} = \mathbf{G}$ say.

But $\overline{CA} = \mathbf{a} - \mathbf{c}$

i.e. $\overline{CA} = -\mathbf{i}_1 - \mathbf{i}_2 - 7\mathbf{i}_3$

therefore

$$\mathbf{G} = \overline{CA} \times \mathbf{F}$$

$$= \frac{-35 + 147}{\sqrt{38}}\mathbf{i}_1 + \frac{-98 + 35}{\sqrt{38}}\mathbf{i}_2 + \frac{-21 + 14}{\sqrt{38}}\mathbf{i}_3 \text{ lb wt. ft units}$$

$$= \frac{112}{\sqrt{38}}\mathbf{i}_1 - \frac{63}{\sqrt{38}}\mathbf{i}_2 - \frac{7}{\sqrt{38}}\mathbf{i}_3 \text{ lb wt. ft units}$$

i.e. $\mathbf{G} = 7(16\mathbf{i}_1 - 9\mathbf{i}_2 - \mathbf{i}_3)/\sqrt{38}$ lb wt. ft units

$$\begin{vmatrix} \mathbf{i}_1 & \mathbf{i}_2 & \mathbf{i}_3 \\ -1 & -1 & -7 \\ 14/\sqrt{38} & 21/\sqrt{38} & 35/\sqrt{38} \end{vmatrix}$$

Examples IIa

Find the scalar products of the following pairs of vectors:

1. $-4\mathbf{i}_1 + 5\mathbf{i}_2 + 3\mathbf{i}_3$ and $2\mathbf{i}_1 + 7\mathbf{i}_2 - 8\mathbf{i}_3$.
2. $3\mathbf{i}_1 + 9\mathbf{i}_2 - 2\mathbf{i}_3$ and $\mathbf{i}_1 - \mathbf{i}_2 - 4\mathbf{i}_3$.
3. $5\mathbf{i}_1 + \mathbf{i}_2 + 2\mathbf{i}_3$ and $-2\mathbf{i}_1 + \mathbf{i}_2 + 3\mathbf{i}_3$.
4. $2\mathbf{i}_1 - 3\mathbf{i}_2 + 6\mathbf{i}_3$ and $2\mathbf{i}_1 - 3\mathbf{i}_2 - 5\mathbf{i}_3$.

Find the vector products of the following pairs of vectors:

5. $-2\mathbf{i}_1 - \mathbf{i}_2 - 3\mathbf{i}_3$ and $4\mathbf{i}_1 + 7\mathbf{i}_2 + 2\mathbf{i}_3$.
6. $2\mathbf{i}_1 - 3\mathbf{i}_2 + 5\mathbf{i}_3$ and $\mathbf{i}_1 - 2\mathbf{i}_2 - 3\mathbf{i}_3$.
7. $\mathbf{i}_1 + 3\mathbf{i}_2 - 8\mathbf{i}_3$ and $-3\mathbf{i}_1 - 5\mathbf{i}_2 + 4\mathbf{i}_3$.
8. $4\mathbf{i}_1 + \mathbf{i}_2 - \mathbf{i}_3$ and $5\mathbf{i}_1 + 2\mathbf{i}_2 - 7\mathbf{i}_3$.

9. Find the work done by a force of 15 lb wt. acting in the direction of the displacement $\overline{AB}$, if its point of application moves from A to C where A, B, C are the points (1, 3, 5), (2, 1, 3), (3, 4, 7) respectively, the length measurements being in feet.

10. A force of 21 dyn acts along the line AB, where A, B are the points (1, 1, 3), (3, 4, 9) respectively. Find the moment of the force about the point (2, 2, 2), distances being measured in centimetres.

11. Find (i) the scalar product, (ii) the vector product, of $\overline{AB}$, $\overline{BC}$ where the position vectors of A, B, C are $\mathbf{i}_1 + 3\mathbf{i}_3$, $5\mathbf{i}_1 + 6\mathbf{i}_2 + 2\mathbf{i}_3$, $7\mathbf{i}_1 + \mathbf{i}_2 + 3\mathbf{i}_3$ respectively.

12. Find (i) the scalar product, (ii) the vector product, of $\overline{AB}$, $\overline{CD}$ where A, B, C, D are the points (1, −1, 1), (2, 3, 4), (1, 3, 5), (2, 6, −4) respectively.

13. A force of 2 lb wt. acts in the direction of $\overline{AB}$ where A, B are the points (2, 4, 3), (5, 5, 4). The point of application moves from A to the point C (6, 4, 4). Find the work done by the force of 2 lb wt., assuming the length measurements to be in feet.

14. A force of 6 lb wt. acts along the line AB, in the direction of the vector $\overline{AB}$, where A, B are the points (2, 3, 6), (1, 1, 8). Find the moment of this force about C (0, 0, -1), the distances being measured in feet.

15. Calculate the angles of triangle ABC where A, B, C are the points (1, 2, 4), (3, 1, -2), (4, 3, 1) respectively.

16. $\mathbf{F}_1, \mathbf{F}_2, \mathbf{F}_3 \ldots \mathbf{F}_n$ are n coplanar forces localized at the points $A_1, A_2, A_3 \ldots A_n$ whose position vectors relative to an origin O are $\mathbf{a}_1, \mathbf{a}_2, \mathbf{a}_3 \ldots \mathbf{a}_n$ respectively. Show that the line of action of the resultant of $\mathbf{F}_1, \mathbf{F}_2, \mathbf{F}_3 \ldots \mathbf{F}_n$ cuts the line through O in the direction of the unit vector $\mathbf{i}$ at a distance r from O where

$$r\mathbf{i} \times (\mathbf{F}_1 + \mathbf{F}_2 + \mathbf{F}_3 + \ldots + \mathbf{F}_n) = \mathbf{a}_1 \times \mathbf{F}_1 + \mathbf{a}_2 \times \mathbf{F}_2 + \\ + \mathbf{a}_3 \times \mathbf{F}_3 + \ldots + \mathbf{a}_n \times \mathbf{F}_n.$$

$ABCD$ is a square of side 6 cm. Forces 6, 4, 3, $4\sqrt{2}$, $4\sqrt{2}$ dyn act along AB, AD, DC, BD, AC respectively, in the directions indicated by the order of the letters. Show that the line of action of their resultant cuts AB at a point $\frac{1}{2}$ cm from A, and find the distance from B of the point at which this line of action cuts BC.

17. If $\mathbf{a}, \mathbf{b}, \mathbf{c}$ are the position vectors of the vertices of triangle ABC, show that

$$\mathbf{b} \times \mathbf{c} + \mathbf{c} \times \mathbf{a} + \mathbf{a} \times \mathbf{b} = 2\Delta\mathbf{w}$$

where Δ is the area of triangle ABC and $\mathbf{w}$ is the unit vector perpendicular to the plane of triangle ABC.

§2.14. Vector Area

If $\overline{OA}$, $\overline{OB}$ represent the vectors $\mathbf{a}$, $\mathbf{b}$ and θ is the angle made by $\overline{OB}$ with $\overline{OA}$, and if $\mathbf{n}$ is the unit vector perpendicular to $\mathbf{a}$ and $\mathbf{b}$ such that $\mathbf{a}, \mathbf{b}, \mathbf{n}$ form a right-handed system, then by definition,

$$\mathbf{a} \times \mathbf{b} = (ab \sin \theta)\mathbf{n}$$

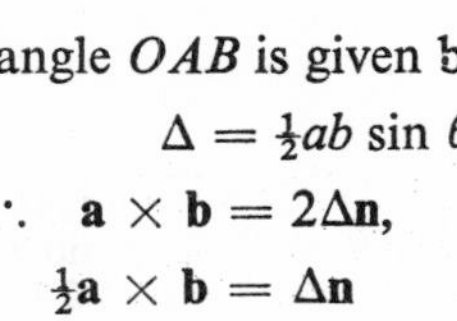

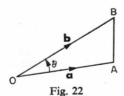

Fig. 22

But the area, Δ, of triangle OAB is given by

$$\Delta = \tfrac{1}{2}ab \sin \theta$$

$$\therefore \quad \mathbf{a} \times \mathbf{b} = 2\Delta\mathbf{n},$$

or

$$\tfrac{1}{2}\mathbf{a} \times \mathbf{b} = \Delta\mathbf{n}$$

Thus the vector $\frac{1}{2}\mathbf{a} \times \mathbf{b}$ has modulus Δ and direction normal to the plane of triangle OAB. This vector, $\frac{1}{2}\mathbf{a} \times \mathbf{b}$, is defined to be the *vector area* of triangle OAB.

§2.15. Triple Products of Three Vectors

If **a**, **b**, **c** are three vectors, any pair of them may be multiplied vectorially to form a new vector **d**; the third of the original vectors may then be multiplied by **d**, either scalarly to form what is known as the *scalar triple product*, or vectorially to form what is known as the *vector triple product*.

§2.16. Scalar Triple Product

Suppose **a**, **b**, **c** are three non-coplanar vectors, which when taken in cyclic order, form a right-handed system. Let **a**, **b**, **c** be represented by $\overline{OA}$, $\overline{OB}$, $\overline{OC}$ respectively.

Complete the parallelepiped $OBDCAB'D'C'$, where OA, BB', DD', CC' are parallel edges. Then if Δ is the area of triangle OBC and α $(=2\Delta)$ is the area of parallelogram $OBDC$, by §2.14,

$$\mathbf{b} \times \mathbf{c} = 2\Delta\mathbf{n}_1$$

i.e.

$$\mathbf{b} \times \mathbf{c} = \alpha\mathbf{n}_1$$

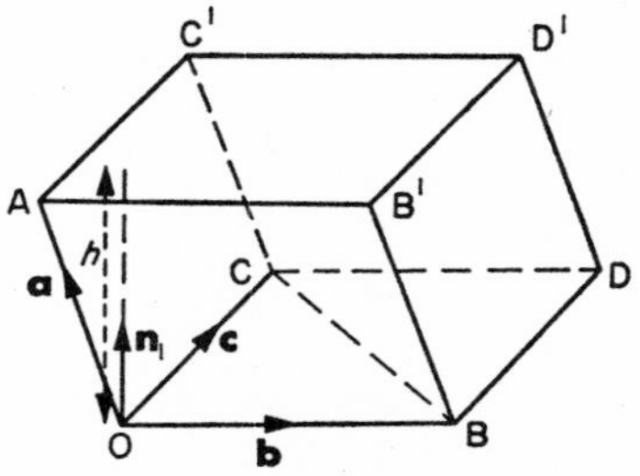

Fig. 23

where $\mathbf{n}_1$ is the unit vector perpendicular to the plane of **b**, **c** such that **b**, **c**, $\mathbf{n}_1$ form a right-handed system. If θ_1 is the angle between **a** and $\mathbf{n}_1$,

$$\begin{aligned} \mathbf{a} \cdot (\mathbf{b} \times \mathbf{c}) &= \mathbf{a} \cdot \alpha\mathbf{n}_1 \\ &= \alpha(\mathbf{a} \cdot \mathbf{n}_1) \\ &= \alpha(a \cos \theta_1) \\ &= \alpha h \\ &= V, \end{aligned}$$

where h $(= a \cos \theta_1)$ is the altitude and V the volume of parallelepiped $OBDC\ AB'\ D'\ C'$, having $OBCD$ as base. By a similar argument it can be shown that

$$\mathbf{b} \cdot (\mathbf{c} \times \mathbf{a}) = \mathbf{c} \cdot (\mathbf{a} \times \mathbf{b}) = V.$$

Hence $\mathbf{a} \cdot (\mathbf{b} \times \mathbf{c}) = \mathbf{b} \cdot (\mathbf{c} \times \mathbf{a}) = \mathbf{c} \cdot (\mathbf{a} \times \mathbf{b}) = V$ (I)

Since $\mathbf{a} \cdot (\mathbf{b} \times \mathbf{c}) = (\mathbf{b} \times \mathbf{c}) \cdot \mathbf{a}$

and $\mathbf{a} \cdot (\mathbf{b} \times \mathbf{c}) = \mathbf{b} \cdot (\mathbf{c} \times \mathbf{a})$

from (I) above,

$$\left.\begin{aligned} &\text{therefore} & \mathbf{b} \cdot (\mathbf{c} \times \mathbf{a}) &= (\mathbf{b} \times \mathbf{c}) \cdot \mathbf{a}. \\ &\text{Similarly} & \mathbf{c} \cdot (\mathbf{a} \times \mathbf{b}) &= (\mathbf{c} \times \mathbf{a}) \cdot \mathbf{b}, \\ &\text{and} & \mathbf{a} \cdot (\mathbf{b} \times \mathbf{c}) &= (\mathbf{a} \times \mathbf{b}) \cdot \mathbf{c}. \end{aligned}\right\} \quad \text{(II)}$$

Further, since $\mathbf{b} \times \mathbf{c} = -(\mathbf{c} \times \mathbf{b})$,

$$\left.\begin{aligned} &\text{therefore} & \mathbf{a} \cdot (\mathbf{b} \times \mathbf{c}) &= -\mathbf{a} \cdot (\mathbf{c} \times \mathbf{b}). \\ &\text{Similarly} & \mathbf{b} \cdot (\mathbf{c} \times \mathbf{a}) &= -\mathbf{b} \cdot (\mathbf{a} \times \mathbf{c}), \\ &\text{and} & \mathbf{c} \cdot (\mathbf{a} \times \mathbf{b}) &= -\mathbf{c} \cdot (\mathbf{b} \times \mathbf{a}). \end{aligned}\right\} \quad \text{(III)}$$

From equations (I) it follows that the value of the scalar triple product is unaltered by changing the order of the letters, provided cyclic order is maintained.

From equations (II) it follows that the value of the scalar triple product is unaltered if the dot and cross are interchanged.

From equations (III) it follows that if one pair of letters in a scalar triple product is interchanged, thus reversing the cyclic order, the sign of the result is changed. It is usual to denote the scalar triple product of **a**, **b**, **c** by [**a b c**]. Equations III can then be expressed more shortly in the form [**a c b**] = −[**a b c**].

It follows that if **a**, **b**, **c** are coplanar, $V = 0$, and therefore [**a b c**] = 0.

§2.17

Suppose that with the usual notation

$$\mathbf{a} = a_1\mathbf{i}_1 + a_2\mathbf{i}_2 + a_3\mathbf{i}_3,$$
$$\mathbf{b} = b_1\mathbf{i}_1 + b_2\mathbf{i}_2 + b_3\mathbf{i}_3,$$
$$\mathbf{c} = c_1\mathbf{i}_1 + c_2\mathbf{i}_2 + c_3\mathbf{i}_3.$$

Then

$$\mathbf{b} \times \mathbf{c} = (b_2c_3 - b_3c_2)\mathbf{i}_1 + (b_3c_1 - b_1c_3)\mathbf{i}_2 + (b_1c_2 - b_2c_1)\mathbf{i}_3,$$

therefore

$$\begin{aligned}\mathbf{a} \,.\, (\mathbf{b} \times \mathbf{c}) &= (a_1\mathbf{i}_1 + a_2\mathbf{i}_2 + a_3\mathbf{i}_3) \,.\, \{(b_2c_3 - b_3c_2)\mathbf{i}_1 \\ &\qquad + (b_3c_1 - b_1c_3)\mathbf{i}_2 + (b_1c_2 - b_2c_1)\mathbf{i}_3)\} \\ &= a_1(b_2c_3 - b_3c_2) + a_2(b_3c_1 - b_1c_3) + a_3(b_1c_2 - b_2c_1)\end{aligned}$$

i.e.

$$[\mathbf{a\ b\ c}] = \begin{vmatrix} a_1 & a_2 & a_3 \\ b_1 & b_2 & b_3 \\ c_1 & c_2 & c_3 \end{vmatrix}$$

§2.18. Vector Triple Product

Suppose vectors **a**, **b**, **c**, $\mathbf{n}_1$ are defined as in §2.16, and suppose $\mathbf{n}_2$ is the unit vector in the direction of **b** while $\mathbf{n}_3$ is the unit vector in the plane of **b**, **c** such that $\mathbf{n}_1$, $\mathbf{n}_2$, $\mathbf{n}_3$ form a mutually perpendicular right-handed system of unit vectors. Then if Δ is the area of triangle OBC (see Fig. 23) and **w** denotes the vector product $\mathbf{a} \times (\mathbf{b} \times \mathbf{c})$,

$$\begin{aligned}\mathbf{w} &= \mathbf{a} \times (\mathbf{b} \times \mathbf{c}) \\ &= \mathbf{a} \times (2\Delta\mathbf{n}_1) \\ &= 2\Delta(\mathbf{a} \times \mathbf{n}_1)\end{aligned}$$

Since $\mathbf{a} \times \mathbf{n}_1$ is a vector perpendicular to the plane of $\mathbf{a}$ and $\mathbf{n}_1$, then $\mathbf{w}$ is a vector perpendicular to $\mathbf{a}$ and $\mathbf{n}_1$, i.e. $\mathbf{w}$ is a vector in the plane of $\mathbf{b}$ and $\mathbf{c}$, perpendicular to $\mathbf{a}$. Thus $\mathbf{w}$ may be expressed in the form

$$\mathbf{w} = \lambda\mathbf{b} + \mu\mathbf{c},$$

where λ, μ are real numbers.

Suppose $\mathbf{a}$, $\mathbf{b}$, $\mathbf{c}$ are expressed in terms of their components in the directions of $\mathbf{n}_1$, $\mathbf{n}_2$, $\mathbf{n}_3$ in the forms

$$\mathbf{a} = a_1\mathbf{n}_1 + a_2\mathbf{n}_2 + a_3\mathbf{n}_3$$
$$\mathbf{b} = b\mathbf{n}_2,$$
$$\mathbf{c} = c_2\mathbf{n}_2 + c_3\mathbf{n}_3$$

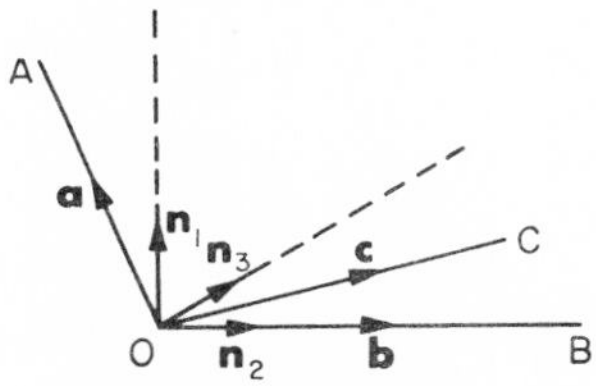

Fig. 24

Then $\qquad \mathbf{b} \times \mathbf{c} = b\mathbf{n}_2 \times (c_2\mathbf{n}_2 + c_3\mathbf{n}_3) = bc_3\mathbf{n}_1$

Hence
$$\begin{aligned} \mathbf{w} &= \mathbf{a} \times (\mathbf{b} \times \mathbf{c}) \\ &= (a_1\mathbf{n}_1 + a_2\mathbf{n}_2 + a_3\mathbf{n}_3) \times bc_3\mathbf{n}_1 \\ &= -a_2bc_3\mathbf{n}_3 + a_3bc_3\mathbf{n}_2 \end{aligned}$$

i.e.
$$\begin{aligned} \mathbf{w} &= (a_3bc_3 + a_2bc_2)\mathbf{n}_2 - (a_2bc_2\mathbf{n}_2 + a_2bc_3\mathbf{n}_3) \\ &= (a_2c_2 + a_3c_3)b\mathbf{n}_2 - (a_2b)(c_2\mathbf{n}_2 + c_3\mathbf{n}_3) \\ &= (\mathbf{a} \cdot \mathbf{c})\mathbf{b} - (\mathbf{a} \cdot \mathbf{b})\mathbf{c} \end{aligned}$$

Thus $\qquad \lambda = \mathbf{a} \cdot \mathbf{c} \quad \text{and} \quad \mu = -(\mathbf{a} \cdot \mathbf{b}),$

therefore $\qquad \mathbf{a} \times (\mathbf{b} \times \mathbf{c}) = (\mathbf{a} \cdot \mathbf{c})\mathbf{b} - (\mathbf{a} \cdot \mathbf{b})\mathbf{c}$

Examples IIb

1. Find the scalar triple product [a b c] in the following cases:

(i) $\mathbf{a} = \mathbf{i}_1 + 3\mathbf{i}_2 + 5\mathbf{i}_3$, $\mathbf{b} = 2\mathbf{i}_1 - \mathbf{i}_2 + 7\mathbf{i}_3$, $\mathbf{c} = 3\mathbf{i}_1 + 2\mathbf{i}_2 - \mathbf{i}_3$;

(ii) $\mathbf{a} = -3\mathbf{i}_1 + 2\mathbf{i}_2 + 6\mathbf{i}_3$, $\mathbf{b} = 4\mathbf{i}_1 + \mathbf{i}_2 + 2\mathbf{i}_3$, $\mathbf{c} = 5\mathbf{i}_1 + 4\mathbf{i}_2 - 2\mathbf{i}_3$;

(iii) $\mathbf{a} = 4\mathbf{i}_1 + 2\mathbf{i}_2 - 3\mathbf{i}_3$, $\mathbf{b} = \mathbf{i}_1 - 2\mathbf{i}_2 + 3\mathbf{i}_3$, $\mathbf{c} = 5\mathbf{i}_1 + 6\mathbf{i}_2 + 7\mathbf{i}_3$;
(iv) $\mathbf{a} = 8\mathbf{i}_1 - 3\mathbf{i}_2 + 3\mathbf{i}_3$, $\mathbf{b} = 4\mathbf{i}_1 + 2\mathbf{i}_2 + 3\mathbf{i}_3$, $\mathbf{c} = 3\mathbf{i}_1 - 4\mathbf{i}_2 - \mathbf{i}_3$;
(v) $\mathbf{a} = 5\mathbf{i}_1 + 3\mathbf{i}_2 - 2\mathbf{i}_3$, $\mathbf{b} = 2\mathbf{i}_1 - 3\mathbf{i}_2 - 3\mathbf{i}_3$, $\mathbf{c} = 3\mathbf{i}_1 + 2\mathbf{i}_2 - 6\mathbf{i}_3$.

2. Find the vector triple product $\mathbf{a} \times (\mathbf{b} \times \mathbf{c})$ in each of the cases (i) to (v) in question **1.**

3. Find the scalar triple product $[\mathbf{p}\,\mathbf{q}\,\mathbf{r}]$ and the vector triple product $\mathbf{q} \times (\mathbf{r} \times \mathbf{p})$ in each of the following cases:

(i) $\mathbf{p} = 2\mathbf{i}_1 + 3\mathbf{i}_2 + 3\mathbf{i}_3$, $\mathbf{q} = 3\mathbf{i}_1 - 2\mathbf{i}_2 - \mathbf{i}_3$, $\mathbf{r} = 2\mathbf{i}_1 + 5\mathbf{i}_2 - 4\mathbf{i}_3$;
(ii) $\mathbf{p} = 2\mathbf{i}_1 + 3\mathbf{i}_2$, $\mathbf{q} = 3\mathbf{i}_2 + 4\mathbf{i}_3$, $\mathbf{r} = 4\mathbf{i}_1 + 5\mathbf{i}_2 + 6\mathbf{i}_3$;
(iii) $\mathbf{p} = \mathbf{i}_1 - 3\mathbf{i}_2 + 4\mathbf{i}_3$, $\mathbf{q} = 2\mathbf{i}_1 + 4\mathbf{i}_2 - \mathbf{i}_3$, $\mathbf{r} = 3\mathbf{i}_1 + 2\mathbf{i}_2 - 3\mathbf{i}_3$;
(iv) $\mathbf{p} = 5\mathbf{i}_2 - \mathbf{i}_3$, $\mathbf{q} = 2\mathbf{i}_1 + 3\mathbf{i}_3$, $\mathbf{r} = 4\mathbf{i}_1 - 3\mathbf{i}_2 - 2\mathbf{i}_3$;
(v) $\mathbf{p} = 3\mathbf{i}_1 + 4\mathbf{i}_2 + 5\mathbf{i}_3$, $\mathbf{q} = 7\mathbf{i}_1 + 24\mathbf{i}_2 + 25\mathbf{i}_3$, $\mathbf{r} = 5\mathbf{i}_1 + 13\mathbf{i}_2 + 12\mathbf{i}_3$.

4. Prove the formula

$$\mathbf{a} \times (\mathbf{b} \times \mathbf{c}) = (\mathbf{a} \cdot \mathbf{c})\mathbf{b} - (\mathbf{a} \cdot \mathbf{b})\mathbf{c},$$

and verify in the case where $\mathbf{a} = \mathbf{i}_1 + 2\mathbf{i}_2 + 3\mathbf{i}_3$, $\mathbf{b} = 3\mathbf{i}_1 + 4\mathbf{i}_2 - \mathbf{i}_3$, $\mathbf{c} = 2\mathbf{i}_1 - 3\mathbf{i}_2 + 4\mathbf{i}_3$.

5. Prove that if $\mathbf{a}$, $\mathbf{b}$, $\mathbf{c}$ are three non-zero vectors, and $(\mathbf{a} \times \mathbf{b}) \times \mathbf{c} = \mathbf{a} \times (\mathbf{b} \times \mathbf{c})$, then $(\mathbf{a} \times \mathbf{c}) \times \mathbf{b}$ is zero.
Discuss the geometrical significance of this result.

6. Verify the formula

$$\mathbf{a} \times (\mathbf{b} \times \mathbf{c}) = (\mathbf{a} \cdot \mathbf{c})\mathbf{b} - (\mathbf{a} \cdot \mathbf{b})\mathbf{c},$$

in the case where

$$\mathbf{a} = 2\mathbf{i}_1 + \mathbf{i}_2,\ \mathbf{b} = 3\mathbf{i}_1 - 2\mathbf{i}_2 + \mathbf{i}_3,\ \mathbf{c} = 2\mathbf{i}_1 - 2\mathbf{i}_2 + \mathbf{i}_3.$$

7. If $\mathbf{p}$, $\mathbf{q}$, $\mathbf{r}$, $\mathbf{s}$ are any four vectors, prove that

$$(\mathbf{p} \times \mathbf{q}) \cdot (\mathbf{r} \times \mathbf{s}) = (\mathbf{p} \cdot \mathbf{r})(\mathbf{q} \cdot \mathbf{s}) - (\mathbf{q} \cdot \mathbf{r})(\mathbf{p} \cdot \mathbf{s}).$$

CHAPTER III

§3.1

Suppose the vector $\mathbf{r}$ is a continuous single-valued function of the scalar variable t. Suppose that when t increases by a small scalar quantity δt, $\mathbf{r}$ increases by a small vector quantity $\delta\mathbf{r}$. Then the derivative of $\mathbf{r}$ with respect to t is defined to be

$$\frac{d\mathbf{r}}{dt} = \lim_{\delta t \to 0} \frac{\delta \mathbf{r}}{\delta t}.$$

Thus the existence of the derivative depends upon the existence of the limit

$$\lim_{\delta t \to 0} \frac{\delta \mathbf{r}}{\delta t}$$

The derivative of $\mathbf{r}$, when it exists, is in general also a function of t, and if $d\mathbf{r}/dt$ is differentiable its derivative is the *second derivative* of $\mathbf{r}$ with respect to t and is written $d^2\mathbf{r}/dt^2$. Similarly for the third, fourth, ... derivatives which are written

$$\frac{d^3\mathbf{r}}{dt^3}, \frac{d^4\mathbf{r}}{dt^4}, \ldots .$$

If x, y, z are the components of $\mathbf{r}$ in the directions of the fixed coordinate axes Ox, Oy, Oz, we may write

$$\mathbf{r} = x\mathbf{i}_1 + y\mathbf{i}_2 + z\mathbf{i}_3,$$

where the scalar quantities x, y, z are functions of t.

Then $$\mathbf{r} + \delta\mathbf{r} = (x + \delta x)\mathbf{i}_1 + (y + \delta y)\mathbf{i}_2 + (z + \delta z)\mathbf{i}_3.$$

Therefore $\qquad \delta\mathbf{r} = \delta x\mathbf{i}_1 + \delta y\mathbf{i}_2 + \delta z\mathbf{i}_3$

and $$\frac{\delta\mathbf{r}}{\delta t} = \frac{\delta x}{\delta t}\mathbf{i}_1 + \frac{\delta y}{\delta t}\mathbf{i}_2 + \frac{\delta z}{\delta t}\mathbf{i}_3$$

and in the limit as $\delta t \to 0$ this becomes

$$\frac{d\mathbf{r}}{dt} = \frac{dx}{dt}\mathbf{i}_1 + \frac{dy}{dt}\mathbf{i}_2 + \frac{dz}{dt}\mathbf{i}_3$$

Similarly

$$\frac{d^2\mathbf{r}}{dt^2} = \frac{d^2x}{dt^2}\mathbf{i}_1 + \frac{d^2y}{dt^2}\mathbf{i}_2 + \frac{d^2z}{dt^2}\mathbf{i}_3, \text{ etc.}$$

Example. Suppose

$$\mathbf{r} = (3t^2 + 2t)\mathbf{i}_1 + (\sin 3t)\mathbf{i}_2 + e^{4t}\mathbf{i}_3,$$

then $$\frac{d\mathbf{r}}{dt} = (6t + 2)\mathbf{i}_1 + (3\cos 3t)\mathbf{i}_2 + 4e^{4t}\mathbf{i}_3,$$

and $$\frac{d^2\mathbf{r}}{dt^2} = 6\mathbf{i}_1 - (9\sin 3t)\mathbf{i}_2 + 16e^{4t}\mathbf{i}_3.$$

§3.2. Derivative of a Sum of Two Vectors

Suppose the vectors $\mathbf{r}_1$ and $\mathbf{r}_2$ are differentiable functions of the scalar variable t, and suppose

$$\mathbf{r} = \mathbf{r}_1 + \mathbf{r}_2$$

Then $$\mathbf{r} + \delta\mathbf{r} = (\mathbf{r}_1 + \delta\mathbf{r}_1) + (\mathbf{r}_2 + \delta\mathbf{r}_2),$$

and $$\delta\mathbf{r} = \delta\mathbf{r}_1 + \delta\mathbf{r}_2$$

Hence $$\frac{\delta\mathbf{r}}{\delta t} = \frac{\delta\mathbf{r}_1}{\delta t} + \frac{\delta\mathbf{r}_2}{\delta t}$$

and in the limit as $\delta t \to 0$

$$\frac{d\mathbf{r}}{dt} = \frac{d\mathbf{r}_1}{dt} + \frac{d\mathbf{r}_2}{dt}$$

Thus, as in scalar algebra, *the derivative of the sum of two vectors is equal to the sum of their derivatives.*

§3.3. Derivative of a Product

Suppose the scalar u and the vectors $\mathbf{r}_1$ and $\mathbf{r}_2$ are all differentiable functions of the scalar variable t.

(i) Suppose the vector $\mathbf{r}$ is given by

$$\mathbf{r} = u\mathbf{r}_1.$$

Then with the usual notation,

$$\mathbf{r} + \delta\mathbf{r} = (u + \delta u)(\mathbf{r}_1 + \delta\mathbf{r}_1),$$

$$= u\mathbf{r}_1 + u(\delta\mathbf{r}_1) + (\delta u)\mathbf{r}_1 + (\delta u)(\delta\mathbf{r}_1),$$

therefore $$\delta\mathbf{r} = u(\delta\mathbf{r}_1) + (\delta u)\mathbf{r}_1 + (\delta u)(\delta\mathbf{r}_1),$$

and $$\frac{\delta\mathbf{r}}{\delta t} = u\,\frac{\delta\mathbf{r}_1}{\delta t} + \frac{\delta u}{\delta t}\mathbf{r}_1 + \frac{(\delta u)(\delta\mathbf{r}_1)}{\delta t}$$

In the limit as $\delta t \to 0$ this becomes

$$\frac{d\mathbf{r}}{dt} = u\,\frac{d\mathbf{r}_1}{dt} + \frac{du}{dt}\mathbf{r}_1$$

(ii) Suppose the scalar r is given by

$$r = \mathbf{r}_1 \,.\, \mathbf{r}_2$$

Then

$$r + \delta r = (\mathbf{r}_1 + \delta\mathbf{r}_1) \,.\, (\mathbf{r}_2 + \delta\mathbf{r}_2),$$

i.e. $$r + \delta r = \mathbf{r}_1 \,.\, \mathbf{r}_2 + \mathbf{r}_1 \,.\, (\delta\mathbf{r}_2) + (\delta\mathbf{r}_1) \,.\, \mathbf{r}_2 + (\delta\mathbf{r}_1) \,.\, (\delta\mathbf{r}_2)$$

Hence $$\delta r = \mathbf{r}_1 \,.\, (\delta\mathbf{r}_2) + (\delta\mathbf{r}_1) \,.\, \mathbf{r}_2 + (\delta\mathbf{r}_1) \,.\, (\delta\mathbf{r}_2),$$

and $$\frac{\delta r}{\delta t} = \mathbf{r}_1 \,.\, \frac{\delta\mathbf{r}_2}{\delta t} + \frac{\delta\mathbf{r}_1}{\delta t} \,.\, \mathbf{r}_2 + \frac{(\delta\mathbf{r}_1) \,.\, (\delta\mathbf{r}_2)}{\delta t}$$

In the limit as $\delta t \to 0$ this becomes

$$\frac{dr}{dt} = \mathbf{r}_1 \,.\, \frac{d\mathbf{r}_2}{dt} + \frac{d\mathbf{r}_1}{dt} \,.\, \mathbf{r}_2$$

(iii) Suppose the vector $\mathbf{r}$ is given by

$$\mathbf{r} = \mathbf{r}_1 \times \mathbf{r}_2$$

Then

$$\mathbf{r} + \delta\mathbf{r} = (\mathbf{r}_1 + \delta\mathbf{r}_1) \times (\mathbf{r}_2 + \delta\mathbf{r}_2)$$

$$= (\mathbf{r}_1 \times \mathbf{r}_2) + (\mathbf{r}_1 \times \delta\mathbf{r}_2) + (\delta\mathbf{r}_1 \times \mathbf{r}_2) + (\delta\mathbf{r}_1 \times \delta\mathbf{r}_2);$$

therefore $\delta\mathbf{r} = (\mathbf{r}_1 \times \delta\mathbf{r}_2) + (\delta\mathbf{r}_1 \times \mathbf{r}_2) + (\delta\mathbf{r}_1 \times \delta\mathbf{r}_2)$,

and $$\frac{\delta\mathbf{r}}{\delta t} = \left(\mathbf{r}_1 \times \frac{\delta\mathbf{r}_2}{\delta t}\right) + \left(\frac{\delta\mathbf{r}_1}{\delta t} \times \mathbf{r}_2\right) + \left(\frac{\delta\mathbf{r}_1}{\delta t} \times \delta\mathbf{r}_2\right).$$

In the limit as $\delta t \to 0$ this becomes

$$\frac{d\mathbf{r}}{dt} = \mathbf{r}_1 \times \frac{d\mathbf{r}_2}{dt} + \frac{d\mathbf{r}_1}{dt} \times \mathbf{r}_2$$

Thus, in each case, the derivative of a product of two functions, at least one of which is a vector, is formed in the same way as the derivative of a product of two scalar functions, with the restriction that, in the case of the vector product of two vectors, the order of the two functions, $\mathbf{r}_1$ *and* $\mathbf{r}_2$, *must remain unaltered.*

§3.4. Derivative of a Triple Product of Three Vectors

Let $\mathbf{r}_1$, $\mathbf{r}_2$, $\mathbf{r}_3$ be three differentiable vector functions of the scalar variable t.

(i) Let $r = [\mathbf{r}_1\mathbf{r}_2\mathbf{r}_3]$.

Then

$$r = \mathbf{r}_1 \,.\, (\mathbf{r}_2 \times \mathbf{r}_3)$$

therefore, from §3.3 (ii),

$$\frac{dr}{dt} = \mathbf{r}_1 \,.\, \frac{d(\mathbf{r}_2 \times \mathbf{r}_3)}{dt} + \frac{d\mathbf{r}_1}{dt} \,.\, (\mathbf{r}_2 \times \mathbf{r}_3)$$

i.e. $$\frac{dr}{dt} = \mathbf{r}_1 \,.\left(\mathbf{r}_2 \times \frac{d\mathbf{r}_3}{dt}\right) + \mathbf{r}_1 \,.\left(\frac{d\mathbf{r}_2}{dt} \times \mathbf{r}_3\right) + \frac{d\mathbf{r}_1}{dt} \,.\, (\mathbf{r}_2 \times \mathbf{r}_3)$$

i.e. $$\frac{dr}{dt} = \left[\mathbf{r}_1\mathbf{r}_2 \frac{d\mathbf{r}_3}{dt}\right] + \left[\mathbf{r}_1 \frac{d\mathbf{r}_2}{dt} \mathbf{r}_3\right] + \left[\frac{d\mathbf{r}_1}{dt} \mathbf{r}_2\mathbf{r}_3\right].$$

(ii) Let $\mathbf{r} = \mathbf{r}_1 \times (\mathbf{r}_2 \times \mathbf{r}_3)$.
Then from §3.3 (iii)

$$\frac{d\mathbf{r}}{dt} = \mathbf{r}_1 \times \frac{d(\mathbf{r}_2 \times \mathbf{r}_3)}{dt} + \frac{d\mathbf{r}_1}{dt} \times (\mathbf{r}_2 \times \mathbf{r}_3)$$

i.e. $$\frac{d\mathbf{r}}{dt} = \mathbf{r}_1 \times \left(\mathbf{r}_2 \times \frac{d\mathbf{r}_3}{dt} + \frac{d\mathbf{r}_2}{dt} \times \mathbf{r}_3\right) + \frac{d\mathbf{r}_1}{dt} \times (\mathbf{r}_2 \times \mathbf{r}_3)$$

i.e.

$$\frac{d\mathbf{r}}{dt} = \mathbf{r}_1 \times \left(\mathbf{r}_2 \times \frac{d\mathbf{r}_3}{dt}\right) + \mathbf{r}_1 \times \left(\frac{d\mathbf{r}_2}{dt} \times \mathbf{r}_3\right) + \frac{d\mathbf{r}_1}{dt} \times (\mathbf{r}_2 \times \mathbf{r}_3).$$

Thus the derivatives of the scalar and vector triple products of three vector functions are found in the same manner as the derivative of a product of three scalar functions, with the limitations that in the case of a scalar triple product the cyclic order of the three functions in the original product must be maintained, and in the case of a vector triple product the actual order of the functions in the original product must remain unaltered.

Example 1. *If* $\mathbf{a}$ *is a fixed vector, and the vectors* $\mathbf{r}_1, \mathbf{r}_2$ *are functions of the scalar variable t, find the derivatives of* (1) $\mathbf{a} \cdot \mathbf{r}_1$, (2) $\mathbf{a} \times \mathbf{r}_1$, (3) $[\mathbf{a}\mathbf{r}_1\mathbf{r}_2]$, (4) $a\mathbf{r}_1 \cdot \mathbf{r}_2$.

(1) Let $r = \mathbf{a} \cdot \mathbf{r}_1$.

Then $$\frac{dr}{dt} = \mathbf{a} \cdot \frac{d\mathbf{r}_1}{dt},$$

since $\mathbf{a}$ is constant and therefore $\frac{d\mathbf{a}}{dt} = 0$.

(2) Let $\mathbf{r} = \mathbf{a} \times \mathbf{r}_1$.

Then $$\frac{d\mathbf{r}}{dt} = \mathbf{a} \times \frac{d\mathbf{r}_1}{dt}$$

(3) Let $r = [\mathbf{a}\mathbf{r}_1\mathbf{r}_2]$.

Then $$\frac{dr}{dt} = \left[\mathbf{a}\mathbf{r}_1 \frac{d\mathbf{r}_2}{dt}\right] + \left[\mathbf{a} \frac{d\mathbf{r}_1}{dt} \mathbf{r}_2\right]$$

(4) Let $r = a\mathbf{r}_1 \cdot \mathbf{r}_2$.

Then $$\frac{dr}{dt} = a\mathbf{r}_1 \cdot \frac{d\mathbf{r}_2}{dt} + a\frac{d\mathbf{r}_1}{dt} \cdot \mathbf{r}_2$$

Example 2. If $$\mathbf{r}_1 = t^2\mathbf{i}_1 + t^3\mathbf{i}_2 + t\mathbf{i}_3,$$

and $$\mathbf{r}_2 = 2t\mathbf{i}_1 - 3t^2\mathbf{i}_2 + 2t^3\mathbf{i}_3,$$

find the derivative with respect to t of $\mathbf{r}_1 \times \mathbf{r}_2$ *and of* $\mathbf{r}_1 \cdot \mathbf{r}_2$.

Let $\mathbf{r} = \mathbf{r}_1 \times \mathbf{r}_2$.

Then $$\frac{d\mathbf{r}}{dt} = \mathbf{r}_1 \times \frac{d\mathbf{r}_2}{dt} + \frac{d\mathbf{r}_1}{dt} \times \mathbf{r}_2$$

$$= (t^2\mathbf{i}_1 + t^3\mathbf{i}_2 + t\mathbf{i}_3) \times (2\mathbf{i}_1 - 6t\mathbf{i}_2 + 6t^2\mathbf{i}_3)$$

$$+ (2t\mathbf{i}_1 + 3t^2\mathbf{i}_2 + \mathbf{i}_3) \times (2t\mathbf{i}_1 - 3t^2\mathbf{i}_2 + 2t^3\mathbf{i}_3)$$

$$= 6(t^5 + t^2)\mathbf{i}_1 - (6t^4 - 2t)\mathbf{i}_2 - 8t^3\mathbf{i}_3$$

$$+ (6t^5 + 3t^2)\mathbf{i}_1 - (4t^4 - 2t)\mathbf{i}_2 - 12t^3\mathbf{i}_3$$

i.e. $$\frac{d\mathbf{r}}{dt} = 3t^2(4t^3 + 3)\mathbf{i}_1 - 2t(5t^3 - 2)\mathbf{i}_2 - 20t^3\mathbf{i}_3$$

Alternatively, the product $\mathbf{r}_1 \times \mathbf{r}_2$ may be evaluated (see Ch. II, Ex. 2) before being differentiated.

Let $s = \mathbf{r}_1 \cdot \mathbf{r}_2$.

Then $$\frac{ds}{dt} = \mathbf{r}_1 \cdot \frac{d\mathbf{r}_2}{dt} + \frac{d\mathbf{r}_1}{dt} \cdot \mathbf{r}_2$$

i.e. $$\frac{ds}{dt} = (t^2\mathbf{i}_1 + t^3\mathbf{i}_2 + t\mathbf{i}_3) \cdot (2\mathbf{i}_1 - 6t\mathbf{i}_2 + 6t^2\mathbf{i}_3)$$

$$+ (2t\mathbf{i}_1 + 3t^2\mathbf{i}_2 + \mathbf{i}_3) \cdot (2t\mathbf{i}_1 - 3t^2\mathbf{i}_2 + 2t^3\mathbf{i}_3),$$

i.e. $$\frac{ds}{dt} = (2t^2 - 6t^4 + 6t^3) + (4t^2 - 9t^4 + 2t^3),$$

i.e. $$\frac{ds}{dt} = 6t^2 + 8t^3 - 15t^4.$$

Example 3. *If* $\mathbf{r} = e^{3t}\mathbf{a} + e^{4t}\mathbf{b}$ *where* **a**, **b** *are constant vectors, show that*

$$\frac{d^2\mathbf{r}}{dt^2} - 7\frac{d\mathbf{r}}{dt} + 12\mathbf{r} = 0.$$

Since $\mathbf{r} = e^{3t}\mathbf{a} + e^{4t}\mathbf{b},$

Then $\dfrac{d\mathbf{r}}{dt} = 3e^{3t}\mathbf{a} + 4e^{4t}\mathbf{b},$

and $\dfrac{d^2\mathbf{r}}{dt^2} = 9e^{3t}\mathbf{a} + 16e^{4t}\mathbf{b},$

therefore

$$\begin{aligned}\frac{d^2\mathbf{r}}{dt^2} - 7\frac{d\mathbf{r}}{dt} + 12\mathbf{r} &= 9e^{3t}\mathbf{a} + 16e^{4t}\mathbf{b} - 21e^{3t}\mathbf{a} - 28e^{4t}\mathbf{b} \\ &\qquad + 12e^{3t}\mathbf{a} + 12e^{4t}\mathbf{b} \\ &= 0\end{aligned}$$

§3.5. Integration

As in scalar calculus, the process of integration consists of being given a certain vector function **r** of the scalar variable t, and finding another function **R** of t such that $d\mathbf{R}/dt = \mathbf{r}$. When this process can be performed, we may write $\mathbf{R} = \int \mathbf{r}\,dt$.

Suppose the function **R** has been found such that $d\mathbf{R}/dt = \mathbf{r}$; then, if **c** is any constant vector, $d(\mathbf{R} + \mathbf{c})/dt$ is also equal to **r**. Hence, in general, if it is known that $d\mathbf{R}/dt = \mathbf{r}$, it is usual to write

$$\int \mathbf{r}\,dt = \mathbf{R} + \mathbf{c},$$

where **c** is a constant whose value can be determined in a particular problem by the specific conditions which govern that problem.

§3.6

In order to carry out the process of integration, it is necessary to express the function to be integrated in a form which is the recognizable derivative of some other known function,

e.g. $$\int \mathbf{a}\sin 3t\,dt = -\tfrac{1}{3}\mathbf{a}\cos 3t + \mathbf{c}.$$

The following devices adapted from scalar calculus will be found useful:

(i) If $\dfrac{d^2\mathbf{r}}{dt^2} = \mu\mathbf{r}$, where μ is a constant scalar, then

$$2\frac{d^2\mathbf{r}}{dt^2} \cdot \frac{d\mathbf{r}}{dt} = 2\mu \frac{d\mathbf{r}}{dt} \cdot \mathbf{r},$$

i.e.
$$\frac{d\mathbf{r}}{dt} \cdot \frac{d\mathbf{r}}{dt} = \mu\mathbf{r} \cdot \mathbf{r} + c$$

where c is a scalar constant.

Hence
$$\left|\frac{d\mathbf{r}}{dt}\right|^2 = \mu r^2 + c$$

(ii) If

$$\frac{d^2\mathbf{r}}{dt^2} - (m + n)\frac{d\mathbf{r}}{dt} + mn\mathbf{r} = 0,$$

where m, n are scalars of the form $m = c + ik$ and $n = c - ik$ so that both $(m + n)$ and mn are real, then

$$\mathbf{r} = e^{mt}\mathbf{a} + e^{nt}\mathbf{b}$$

where $\mathbf{a}$, $\mathbf{b}$ are constant vectors;

i.e.
$$\mathbf{r} = e^{ct}(e^{ikt}\mathbf{a} + e^{-ikt}\mathbf{b}),$$

i.e.
$$\mathbf{r} = e^{ct}(\mathbf{A}\cos kt + \mathbf{B}\sin kt)$$

where $\mathbf{A}$, $\mathbf{B}$ are constant vectors, and are functions of $\mathbf{a}$, $\mathbf{b}$.

Examples III

1. Differentiate with respect to t:

(i) $(2\sin t)\mathbf{i}_1 + (3\cos t)\mathbf{i}_2 + (\sin 2t)\mathbf{i}_3$;
(ii) $3t^4\mathbf{i}_1 + 5t^3\mathbf{i}_2 + 4t\mathbf{i}_3$;
(iii) $e^{2t}\mathbf{i}_1 + e^{3t}\mathbf{i}_2 + e^{4t}\mathbf{i}_3$;
(iv) $(3t^2 + 5)\mathbf{i}_1 + (t^3 - 4t^2 + 3t)\mathbf{i}_2 + (\sin 4t)\mathbf{i}_3$;
(v) $(t^2 + 1)^{\frac{1}{2}}\mathbf{i}_1 + (\log t)\mathbf{i}_2 + 3t^4\mathbf{i}_3$.

2. If $\mathbf{r}_1 = 2t\mathbf{i}_1 + 3t^2\mathbf{i}_2 + 5t^3\mathbf{i}_3$, and $\mathbf{r}_2 = t^3\mathbf{i}_1 + t\mathbf{i}_2 - t^3\mathbf{i}_3$, and $\mathbf{a} = 2\mathbf{i}_1 - 3\mathbf{i}_2 + 4\mathbf{i}_3$, find the derivatives with respect to t of:

(i) $\mathbf{r}_1 + \mathbf{r}_2$; (ii) $\mathbf{a} \cdot \mathbf{r}_1$; (iii) $\mathbf{a} \times \mathbf{r}_2$;
(iv) $\mathbf{r}_1 \times \mathbf{r}_2$; (v) $(\mathbf{a} + \mathbf{r}_1) \cdot \mathbf{r}_2$; (vi) $[\mathbf{a}\mathbf{r}_1\mathbf{r}_2]$;
(vii) $\mathbf{a} \times (\mathbf{r}_1 \times \mathbf{r}_2)$; (viii) $\mathbf{r}_1 \cdot \mathbf{r}_1$; (ix) $\mathbf{r}_1 \cdot \mathbf{r}_2$;
(x) $\mathbf{r}_2 \cdot \mathbf{r}_2$.

3. If **a**, **b** are constant vectors, and $\mathbf{r} = (\cos 2t)\mathbf{a} + (\sin 2t)\mathbf{b}$ show that

(i) $\dfrac{d^2\mathbf{r}}{dt^2} + 4\mathbf{r} = 0$, and

(ii) $\mathbf{r} \times \dfrac{d\mathbf{r}}{dt} = 2\mathbf{a} \times \mathbf{b}$.

4. Evaluate the following integrals:

(i) $\int (t^2\mathbf{i}_1 + t\mathbf{i}_2 + \mathbf{i}_3)dt$; (ii) $\int \{(\sin t)\mathbf{i}_1 + (\cos t)\mathbf{i}_2 + t\mathbf{i}_3\}dt$;

(iii) $\int \mathbf{a}e^t dt$; (iv) $\int (t\mathbf{i}_1 + t^3\mathbf{i}_2 + \mathbf{i}_3) \,.\, (\mathbf{i}_1 + 2\mathbf{i}_2)dt$.

5. Verify that

$$\int\Big(\{\mathbf{i}_1 + (\cos t)\mathbf{i}_2\} \times \{(\cos t)\mathbf{i}_1 - t\mathbf{i}_2\}$$
$$+ \{t\mathbf{i}_1 + (\sin t)\mathbf{i}_2\} \times \{(-\sin t)\mathbf{i}_1 - \mathbf{i}_2\}\Big)dt$$

is equal to $-\{t^2 + (\sin t)(\cos t)\}\mathbf{i}_3$.

6. Evaluate the following integrals:

(i) $\int e^t(1 + t)\mathbf{a} \,.\, \mathbf{b}\, dt$,

(ii) $\int \{(\sin t)\mathbf{a} + (\cos t)\mathbf{b}\} \times \{(\cos t)\mathbf{a} + (\sin t)\mathbf{b}\}dt$,

where **a**, **b** are constant vectors.

7. Show that if $\mathbf{r}_1 = t\mathbf{a} + \dfrac{1}{t}\mathbf{b}$, and $\mathbf{r}_2 = \dfrac{1}{t}\mathbf{a} + t\mathbf{b}$,

where **a**, **b** are constant vectors, then

$$\frac{d(\mathbf{r}_1 \times \mathbf{r}_2)}{dt} = 2\left(t + \frac{1}{t^3}\right)\mathbf{a} \times \mathbf{b}.$$

8. Differentiate with respect to t the vector $36t^3\mathbf{i}_1 + (\log t)\mathbf{i}_2 + \dfrac{1}{t}\mathbf{i}_3$.

9. Differentiate with respect to t, (i) $\mathbf{r}_1 \times \mathbf{r}_2$, (ii) $\mathbf{r}_1 \,.\, \mathbf{r}_2$, where $\mathbf{r}_1 = t^2\mathbf{i}_1 - t\mathbf{i}_2 + 3\mathbf{i}_3$ and $\mathbf{r}_2 = 3\mathbf{i}_1 + t\mathbf{i}_2 - t^2\mathbf{i}_3$.

CHAPTER IV

§4.1

Suppose O is a fixed origin, and suppose that at time t a moving point is at the point P, whose position vector referred to O is $\mathbf{r}$. Suppose that after a small interval of time δt, the moving point has reached the point P', whose position vector is $\mathbf{r} + \delta\mathbf{r}$, and let the arc PP' be of length δs.

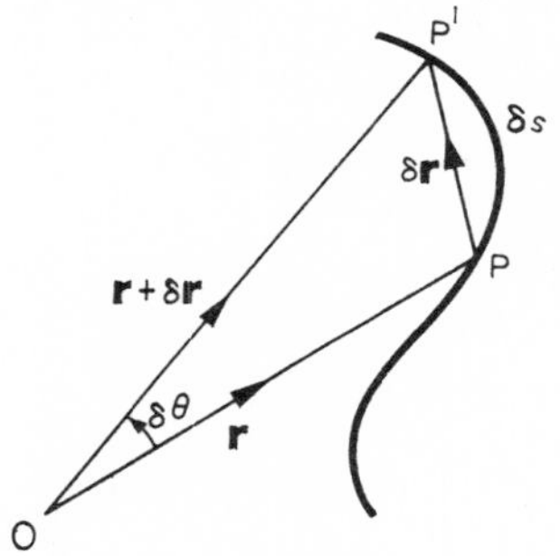

Fig. 25

Then $$\overline{PP'} = \delta\mathbf{r},$$

and $$PP' = |\,\delta\mathbf{r}\,| \simeq \delta s.$$

If $\boldsymbol{\tau}$ is the unit vector along the tangent at P,

$$\delta\mathbf{r} \to \boldsymbol{\tau}(\delta s),$$

and if $\mathbf{v}$ is the velocity of the moving point at time t,

$$\mathbf{v} = \lim_{\delta t \to 0} \frac{\delta\mathbf{r}}{\delta t} = \frac{d\mathbf{r}}{dt} = \dot{\mathbf{r}}$$

Also $$\mathbf{v} = \lim_{\delta t \to 0} \left(\frac{\delta\mathbf{r}}{\delta s}\right)\left(\frac{\delta s}{\delta t}\right) = \boldsymbol{\tau} v$$

where $v = |\mathbf{v}| = \lim_{\delta t \to 0} \frac{\delta s}{\delta t}$, and is the speed of the moving point along its path.

If $\mathbf{f}$ is the acceleration of the moving point,

$$\mathbf{f} = \lim_{\delta t \to 0} \frac{\delta \mathbf{v}}{\delta t} = \frac{d\mathbf{v}}{dt}$$

or

$$\mathbf{f} = \frac{d^2\mathbf{r}}{dt^2} = \ddot{\mathbf{r}}.$$

§4.2

As the moving point moves from P to P', suppose the line joining it to the origin turns through an angle $\delta\theta$, so that $P\hat{O}P' = \delta\theta$. Let Q be a point on OP' such that $OQ = OP = r$.

Then
$$QP' = \delta r,$$
and
$$PQ \simeq r\delta\theta.$$

As $\delta t \to 0$, the lines OP, OP', may be taken to be perpendicular to PQ.

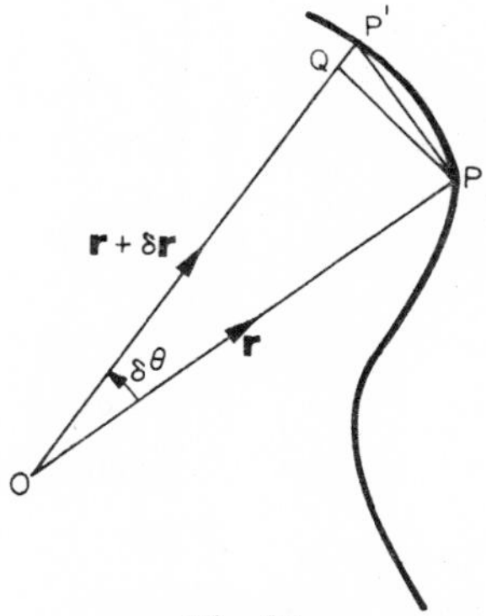

Fig. 26

Let $\mathbf{l}_1$, $\mathbf{l}_2$ be unit vectors along OP and in the *transverse direction* where the transverse direction is defined to be the direction perpendicular to OP in the plane of OP, OP', whose positive sense is that of $\overline{PQ}$. Then when $\delta t \to 0$, $\mathbf{l}_2$ is in the direction $\overline{PQ}$. Let $\mathbf{l}_3$ be a third unit vector, perpendicular to the plane of OP, OP' such that $\mathbf{l}_1$, $\mathbf{l}_2$, $\mathbf{l}_3$ form a mutually perpendicular right-handed system of unit vectors.

Then with the same notation,

$$\mathbf{v} = \lim_{\delta t \to 0} \frac{\delta \mathbf{r}}{\delta t},$$

i.e. $$\mathbf{v} = \lim_{\delta t \to 0} \frac{\overline{PP'}}{\delta t}$$

i.e. $$\mathbf{v} = \lim_{\delta t \to 0} \frac{\overline{PQ} + \overline{QP'}}{\delta t}$$

i.e. $$\mathbf{v} = \lim_{\delta t \to 0} \frac{\overline{PQ}}{\delta t} + \lim_{\delta t \to 0} \frac{\overline{QP'}}{\delta t}$$

i.e. $$\mathbf{v} = r\frac{d\theta}{dt}\mathbf{l}_2 + \frac{dr}{dt}\mathbf{l}_1$$

i.e. $$\mathbf{v} = \dot{\mathbf{r}} = \dot{r}\mathbf{l}_1 + r\dot{\theta}\mathbf{l}_2. \qquad \text{(I)}$$

§4.3

Using the same notation as in the preceding paragraphs, let $\delta\mathbf{l}_1$, $\delta\mathbf{l}_2$ be the small increments in $\mathbf{l}_1$, $\mathbf{l}_2$, as the moving point moves

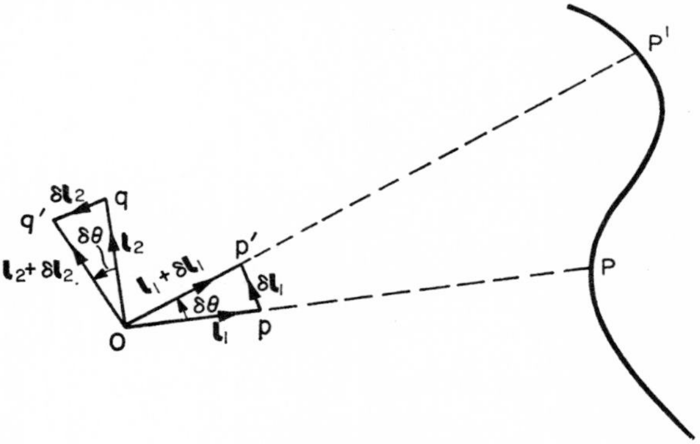

Fig. 27

from P to P', so that $\mathbf{l}_1$, $\mathbf{l}_2$ are in the directions $\overline{OP}$, and the transverse direction while $\mathbf{l}_1 + \delta\mathbf{l}_1$, $\mathbf{l}_2 + \delta\mathbf{l}_2$ are also unit vectors, but in the directions $\overline{OP}'$ and perpendicular to $\overline{OP}'$ respectively. Let unit lengths Op, Op' be cut off on OP, OP', and let lines Oq, Oq' be drawn, of unit length, perpendicular to Op, Op' so that $\overline{Op}$, $\overline{Oq}$, represent $\mathbf{l}_1$, $\mathbf{l}_2$ and $\overline{Op}'$, $\overline{Oq}'$, represent $\mathbf{l}_1 + \delta\mathbf{l}_1$, $\mathbf{l}_2 + \delta\mathbf{l}_2$. Then $\overline{pp}' = \delta\mathbf{l}_1$ and $\overline{qq}' = \delta\mathbf{l}_2$.

But $\quad pp' \simeq Op\,\delta\theta, \quad$ and $\quad qq' \simeq Oq\,\delta\theta,$

i.e. $\quad pp' \simeq \delta\theta, \quad$ and $\quad qq' \simeq \delta\theta.$

Since pp' is in the direction of $\overline{Oq}$, and $\overline{qq}'$ is in the direction of $\overline{pO}$, then as $\delta t \to 0$,

$$\overline{pp}' \to (\delta\theta)\mathbf{l}_2 \quad \text{and} \quad \overline{qq}' \to (\delta\theta)(-\mathbf{l}_1)$$

But $\quad \overline{pp}' = \delta\mathbf{l}_1 \quad$ and $\quad \overline{qq}' = \delta\mathbf{l}_2$

therefore $\quad \delta\mathbf{l}_1 \to (\delta\theta)\mathbf{l}_2, \quad$ and $\quad \delta\mathbf{l}_2 \to -(\delta\theta)\mathbf{l}_1,$

i.e. $\quad \dfrac{d\mathbf{l}_1}{dt} = \dot{\theta}\,\mathbf{l}_2, \quad$ and $\quad \dfrac{d\mathbf{l}_2}{dt} = -\,\dot{\theta}\,\mathbf{l}_1$

§4.4

From §4.2, $\quad \mathbf{v} = \dot{\mathbf{r}} = \dot{r}\mathbf{l}_1 + r\dot{\theta}\mathbf{l}_2$

and from §4.1, $\quad \mathbf{f} = \ddot{\mathbf{r}} = \dfrac{d\mathbf{v}}{dt}$

Hence $\quad \mathbf{f} = \dfrac{d}{dt}(\dot{r}\mathbf{l}_1 + r\dot{\theta}\mathbf{l}_2),$

i.e. $\quad \mathbf{f} = \ddot{r}\,\mathbf{l}_1 + \dot{r}\,\dfrac{d\mathbf{l}_1}{dt} + \dot{r}\dot{\theta}\mathbf{l}_2 + r\ddot{\theta}\mathbf{l}_2 + r\dot{\theta}\,\dfrac{d\mathbf{l}_2}{dt}$

i.e. $\quad \mathbf{f} = \ddot{r}\mathbf{l}_1 + \dot{r}\dot{\theta}\mathbf{l}_2 + \dot{r}\dot{\theta}\mathbf{l}_2 + r\ddot{\theta}\mathbf{l}_2 - r\dot{\theta}^2\mathbf{l}_1$

i.e. $\quad \mathbf{f} = (\ddot{r} - r\dot{\theta}^2)\mathbf{l}_1 + (r\ddot{\theta} + 2\dot{r}\dot{\theta})\mathbf{l}_2.$

Thus the radial component of the acceleration is $(\ddot{r} - r\dot{\theta}^2)$, and the transverse component is $(2\dot{r}\dot{\theta} + r\ddot{\theta})$ or $\dfrac{1}{r}\dfrac{d}{dt}(r^2\dot{\theta})$.

Example 1. *A point moves so that at time t its position vector* **r** *is given by*

$$\mathbf{r} = (\sin 2t)\mathbf{i}_1 + (\cos 3t)\mathbf{i}_2 + e^t\mathbf{i}_3.$$

Find the velocity and acceleration of the point at time t.

Since $\mathbf{r} = (\sin 2t)\mathbf{i}_1 + (\cos 3t)\mathbf{i}_2 + e^t\mathbf{i}_3,$

therefore $\dot{\mathbf{r}} = (2\cos 2t)\mathbf{i}_1 - (3\sin 3t)\mathbf{i}_2 + e^t\mathbf{i}_3,$

and $\ddot{\mathbf{r}} = -(4\sin 2t)\mathbf{i}_1 - (9\cos 3t)\mathbf{i}_2 + e^t\mathbf{i}_3$

Hence the velocity has components $2\cos 2t$, $-3\sin 3t$ and e^t; and the acceleration has components $-4\sin 2t$, $-9\cos 3t$, e^t.

Example 2. *A particle moves on the plane curve $r = a\sin\theta$, where a is a constant number. Find the radial and transverse components of the velocity and acceleration at any time t.*

Since the particle moves in a plane curve, $\dot{\theta}$ and $\ddot{\theta}$ are the angular velocity and acceleration of the line joining the particle to the origin.

The equation of the path of the particle is

$$r = a\sin\theta,$$

therefore $\dot{r} = a(\cos\theta)\dot{\theta},$

and $\ddot{r} = (-a\sin\theta)\dot{\theta}^2 + (a\cos\theta)\ddot{\theta}.$

Hence if **v** is the velocity of the particle

$$\mathbf{v} = (a\cos\theta)\dot{\theta}\mathbf{l}_1 + r\dot{\theta}\mathbf{l}_2$$

i.e. $\mathbf{v} = (a\cos\theta)\dot{\theta}\mathbf{l}_1 + (a\sin\theta)\dot{\theta}\mathbf{l}_2$

and if **f** is the acceleration of the particle,

$$\mathbf{f} = \{(-a\sin\theta)\dot{\theta}^2 + (a\cos\theta)\ddot{\theta} - r\dot{\theta}^2\}\mathbf{l}_1 + \{(2a\cos\theta)\dot{\theta}^2 + r\ddot{\theta}\}\mathbf{l}_2$$

i.e. $\mathbf{f} = a\{(\cos\theta)\ddot{\theta} - 2(\sin\theta)\dot{\theta}^2\}\mathbf{l}_1 + a\{2(\cos\theta)\dot{\theta}^2 + (\sin\theta)\ddot{\theta}\}\mathbf{l}_2$

Hence the radial and transverse components of the velocity are $(a\cos\theta)\dot{\theta}$ and $(a\sin\theta)\dot{\theta}$, while the radial and transverse components of the acceleration are $a\{(\cos\theta)\ddot{\theta} - 2(\sin\theta)\dot{\theta}^2\}$ and $a\{2(\cos\theta)\dot{\theta}^2 + (\sin\theta)\ddot{\theta}\}$ respectively.

Examples IVa

1. Find the velocity and acceleration of a particle which moves so that at time t its position vector is $\mathbf{r}$ where:

(i) $\mathbf{r} = (\sin t)\mathbf{i}_1 + (\cos t)\mathbf{i}_2 + t\mathbf{i}_3$;
(ii) $\mathbf{r} = t^3\mathbf{i}_1 + t^2\mathbf{i}_2 + 3t\mathbf{i}_3$;
(iii) $\mathbf{r} = (\log t)\mathbf{i}_1 - e^{2t}\mathbf{i}_2 + t^2\mathbf{i}_3$;
(iv) $\mathbf{r} = \mathbf{a}e^{2t}$ where $\mathbf{a}$ is a constant vector;
(v) $\mathbf{r} = \mathbf{a}\sin 2t + \mathbf{b}\cos 2t$ where $\mathbf{a}$, $\mathbf{b}$ are constant vectors.

2. Find the velocity and acceleration of a particle which moves along the plane curve $r = ae^{\theta}$ in such a way that the line joining the particle to the origin turns with constant angular velocity ω.

3. Find the radial and transverse components of the acceleration of a particle P which moves in the plane curve $\frac{3}{r} = 1 + \cos\theta$ in such a way that OP rotates with constant angular velocity ω, O being the origin.

4. Find the velocity and acceleration of a particle which moves so that at time t its position vector $\mathbf{r}$ is given by

(i) $\mathbf{r} = a(t - \sin t)\mathbf{i}_1 + a(1 - \cos t)\mathbf{i}_2 + bt\mathbf{i}_3$;
(ii) $\mathbf{r} = a\cos t\mathbf{i}_1 + a\sin t\mathbf{i}_2 + a\cos 2t\mathbf{i}_3$.

5. If a particle moves along the plane curve $r = a\theta$ where $\theta = kt$, find the radial and transverse components of its velocity and acceleration.

6. Find the radial and transverse components of the velocity and acceleration of a particle moving along a plane curve given that

(i) the equation of the curve is $r = a/\theta$, and $\theta = t + 1/t$;
(ii) the equation of the curve is $r^2\theta = 4$, and $\theta = t/2$;
(iii) the equation of the curve is $r = 1 - 1/\theta$, and $\theta = e^t$;
(iv) the equation of the curve is $r = 4 - \cos\theta$, and $\theta = 2t$.

7. Find the radial and transverse components of the velocity and acceleration of a particle moving along a plane curve in such a way that $r^2\dot{\theta} = 1$, given that the equation of the curve is

(i) $r = 6/(2 - \cos\theta)$; (ii) $r = 5/(3 + 2\cos\theta)$.

8. Find the velocity and acceleration of a particle moving along the line

$$\mathbf{r} = \mathbf{a} + p\mathbf{b}$$

where $\mathbf{a}$ and $\mathbf{b}$ are constant vectors, and p is a scalar function of t such that

(i) $p = 3t^2$;
(ii) $p = 4t$;
(iii) $p = \sin t$;
(iv) $p = t + \log t$.

§4.5. Momentum

If at time t a particle of mass m is situated at a point P, whose position vector referred to a fixed origin O is $\mathbf{r}$, its *linear momentum* is defined to be the vector $m\dot{\mathbf{r}}$. The product $\mathbf{r} \times m\dot{\mathbf{r}}$ is called the *moment of momentum*, or *angular momentum* of the particle, and is the moment about O of a vector $m\dot{\mathbf{r}}$ drawn through P. The *linear momentum* is usually called simply the *momentum*.

If $\mathbf{K}$ denotes the linear momentum, and $\mathbf{H}$ the angular momentum, of a moving particle whose position vector at time t is $\mathbf{r}$, then

$$\mathbf{H} = \mathbf{r} \times \mathbf{K}.$$

§4.6. Newton's Laws of Motion Applied to a Particle

I. *Every body continues in its state of rest, or of uniform motion in a straight line, unless it is compelled to change that state by an external impressed force.*

Motion implies direction, and hence that which changes the motion of a body must be a vector, and is known as a *force*.

II. *The rate of change of linear momentum of a body is proportional to the external impressed force and takes place in the same direction.*

Thus, if a particle of mass m, acted upon by a force $\mathbf{F}$ is at a point whose position vector is $\mathbf{r}$ at time t, then

$$\mathbf{F} \propto \frac{d}{dt}(m\dot{\mathbf{r}}),$$

i.e.

$$\mathbf{F} \propto m\ddot{\mathbf{r}}, \text{ if } m \text{ is constant.}$$

If the units are suitably chosen (the *absolute units* of classical mechanics), this may be written

$$\mathbf{F} = m\ddot{\mathbf{r}} = \frac{d\mathbf{K}}{dt}.$$

III. *To every action there is an equal and opposite reaction.*

Thus if a body A acts upon a body B with a force $\mathbf{F}$, then B will act upon A with a force $-\mathbf{F}$.

§4.7

If a force $\mathbf{F}$ acts upon a particle of mass m whose position vector at time t is $\mathbf{r}$, then

$$\mathbf{F} = m\ddot{\mathbf{r}},$$

therefore
$$\mathbf{r} \times \mathbf{F} = \mathbf{r} \times m\ddot{\mathbf{r}}. \qquad (1)$$

Now
$$\mathbf{H} = \mathbf{r} \times m\dot{\mathbf{r}}$$

therefore
$$\frac{d\mathbf{H}}{dt} = \dot{\mathbf{r}} \times m\dot{\mathbf{r}} + \mathbf{r} \times m\ddot{\mathbf{r}}.$$

But
$$\dot{\mathbf{r}} \times m\dot{\mathbf{r}} = 0$$

therefore
$$\frac{d\mathbf{H}}{dt} = \mathbf{r} \times m\ddot{\mathbf{r}}. \qquad (2)$$

Thus, from (1) and (2)

$$\mathbf{r} \times \mathbf{F} = \frac{d\mathbf{H}}{dt}$$

i.e. *the moment, about an arbitrary point O, of the force acting on a particle is equal to the rate of change of angular momentum of the particle about O.* This result is analogous to the statement that *the force acting on a particle is equal to the rate of change of linear momentum of the particle,* and the two statements together make Newton's Second Law generally applicable in Vector Mechanics.

Example 3. *A particle of mass* 2 *lb starts from the origin with a velocity of* 15 *ft/sec along the line whose direction ratios are* 1 : 2 : 2. *The particle is acted upon by a force whose value at any time t is* $(2\mathbf{i}_1 + t\mathbf{i}_2 + t^2\mathbf{i}_3)$ *pdl. where t is measured in sec. Find the position of the particle at time t.*

Let $\mathbf{r}$ be the position vector of the particle at time t. Then

$$2\ddot{\mathbf{r}} = 2\mathbf{i}_1 + t\mathbf{i}_2 + t^2\mathbf{i}_3$$

therefore
$$2\dot{\mathbf{r}} = 2t\mathbf{i}_1 + \tfrac{1}{2}t^2\mathbf{i}_2 + \tfrac{1}{3}t^3\mathbf{i}_3 + \mathbf{a} \qquad (1)$$

where $\mathbf{a}$ is a constant vector.

But when $t = 0,\ \dot{\mathbf{r}} = 15(\frac{1}{3}\mathbf{i}_1 + \frac{2}{3}\mathbf{i}_2 + \frac{2}{3}\mathbf{i}_3)$ (2)

therefore $30(\frac{1}{3}\mathbf{i}_1 + \frac{2}{3}\mathbf{i}_2 + \frac{2}{3}\mathbf{i}_3) = 0 + \mathbf{a}$

Hence equation (1) may be written

$$2\dot{\mathbf{r}} = (2t + 10)\mathbf{i}_1 + (\tfrac{1}{2}t^2 + 20)\mathbf{i}_2 + (\tfrac{1}{3}t^3 + 20)\mathbf{i}_3$$

therefore $2\mathbf{r} = (t^2 + 10t)\mathbf{i}_1 + (\frac{1}{6}t^3 + 20t)\mathbf{i}_2 + (\frac{1}{12}t^4 + 20t)\mathbf{i}_3 + \mathbf{b},$

where $\mathbf{b}$ is a constant vector.

But $\mathbf{r} = 0$ when $t = 0$, therefore $\mathbf{b} = 0$,

therefore $\mathbf{r} = (\frac{1}{2}t^2 + 5t)\mathbf{i}_1 + (\frac{1}{12}t^3 + 10t)\mathbf{i}_2 + (\frac{1}{24}t^4 + 10t)\mathbf{i}_3.$

Example 4. *A particle of mass* 5 *lb is projected from the point* (2, 3, 6) *with a speed of* 70 *ft/sec. If the particle is at the point whose position vector is* $\mathbf{r}$ *at time* t, *and is acted upon by a force of* $(-5\mathbf{r})$ *pdl, find the speed of the particle when it is* 15 *ft from the origin.*

Since the particle is subject to a force $(-5\mathbf{r})$ pdl, then

$$5\ddot{\mathbf{r}} = -5\mathbf{r},$$

i.e. $\ddot{\mathbf{r}} = -\mathbf{r}.$

Multiplying both sides of this equation scalarly by $2\dot{\mathbf{r}}$,

$$2\dot{\mathbf{r}} \cdot \ddot{\mathbf{r}} = -2\dot{\mathbf{r}} \cdot \mathbf{r},$$

therefore $\dot{\mathbf{r}} \cdot \dot{\mathbf{r}} = -\mathbf{r} \cdot \mathbf{r} + a,$ (1)

where a is a scalar constant.

Equation (1) may be written

$$v^2 = a - r^2 \quad (2)$$

where $\mathbf{v}$ $(= \dot{\mathbf{r}})$ is the velocity of the particle at time t. But $v = 70$ when $r^2 = 2^2 + 3^2 + 6^2 = 49$,

therefore $4900 = a - 49$

$$a = 4949.$$

Then equation (2) may be written

$$v^2 = 4949 - r^2 \quad (3)$$

If v_1 is the speed of the particle when it is 15 ft from the origin, then $v = v_1$ when $r = 15$. Hence from equation (3)

$$v_1^2 = 4949 - 225$$
$$= 4724,$$

therefore $$v = \sqrt{(4724)} \text{ ft/sec}$$

i.e. $$v = 68.73 \text{ ft/sec.}$$

Example 5. *A particle is projected with a velocity of* 60 *ft/sec in a direction making an angle* $\cos^{-1} 4/5$ *with the upward vertical. Find the velocity and position of the particle at time t, assuming that the particle is moving freely under the action of gravity.*

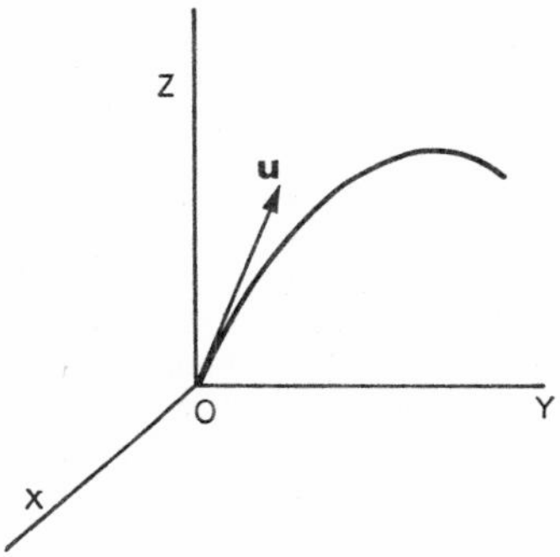

Fig. 28

Let OX, OY, OZ be a mutually perpendicular right-handed system of axes through the point of projection O, such that OZ is along the upward vertical through O, and OY, OZ are in the plane of the initial velocity $\mathbf{u}$. Then, with the usual notation, if $\mathbf{r}$ is the position vector of the particle at time t,

$$\ddot{\mathbf{r}} = -g\mathbf{i}_3,$$

therefore $$\dot{\mathbf{r}} = -gt\mathbf{i}_3 + \mathbf{a},$$

where $\mathbf{a}$ is a constant vector. But when $t = 0$,

$$\dot{\mathbf{r}} = \mathbf{u} = 36\mathbf{i}_2 + 48\mathbf{i}_3$$

therefore $\mathbf{a} = 36\mathbf{i}_2 + 48\mathbf{i}_3.$

Hence $\dot{\mathbf{r}} = -gt\,\mathbf{i}_3 + 36\mathbf{i}_2 + 48\mathbf{i}_3.$ (1)

Integrating again,

$$\mathbf{r} = -\tfrac{1}{2}gt^2\mathbf{i}_3 + (36\mathbf{i}_2 + 48\mathbf{i}_3)t + \mathbf{b},$$

where $\mathbf{b}$ is a constant vector. But $\mathbf{r} = 0$ when $t = 0$,

therefore $\mathbf{b} = 0,$

hence $\mathbf{r} = 36t\mathbf{i}_2 + (48t - \tfrac{1}{2}gt^2)\mathbf{i}_3.$ (2)

Equation (1) may be written

$$\dot{\mathbf{r}} = 36\mathbf{i}_2 + (48 - gt)\mathbf{i}_3. \quad (3)$$

Hence at time t the particle is at the point $(0, 36t, 48t - 16t^2)$ and is moving with velocity $\mathbf{v}$ where $\mathbf{v} = 36\mathbf{i}_2 + (48 - 32t)\mathbf{i}_3$.

Examples IVb

Assume g to be 32 ft/sec², and take the time t to be measured in seconds unless stated otherwise.

1. A particle of mass 10 lb is acted upon by a constant force $\mathbf{F}$, where $\mathbf{F} = (50\mathbf{i}_1 + 25\mathbf{i}_2 + 30\mathbf{i}_3)$ pdl. Initially the particle has a velocity of 36 ft/sec along the line whose direction ratios are 4 : 4 : 7. Find the position and velocity of the particle at any time t.

2. A particle of mass 5 lb is projected from the origin with a velocity of 28 ft/sec along the line whose direction ratios are 2 : 3 : 6. The particle is acted upon by a force $\mathbf{F}$, whose value at time t is given by

$$\mathbf{F} = \{(10 \sin t)\mathbf{i}_1 - 320\mathbf{i}_2\} \text{ pdl.}$$

Find the velocity and position of the particle at time t.

3. A particle is projected from a point O with a velocity of 88 ft/sec along the line whose direction ratios are 2 : 6 : 9. The particle is subject to an acceleration whose value at time t is $\{t\mathbf{i}_1 + (\sin t)\mathbf{i}_2 - 32\mathbf{i}_3\}$ ft/sec². Find the velocity and position of the particle at time t.

4. A particle is projected from the origin with a velocity of 65 ft/sec along a line making an angle $\tan^{-1}5/12$ with the upward vertical. If the particle moves freely under gravity, find its velocity and position at any time t, and show that its trajectory is a curve in a vertical plane. Find also the time at which the particle again reaches the horizontal plane through O, and its distance from O at that instant.

5. A particle is projected with a velocity of 64 ft/sec in a direction making an angle of 30° with the horizontal. Find (i) the time at which the particle reaches the horizontal plane through the point of projection, (ii) the distance at that time from the point of projection, (iii) the greatest height attained during the motion.

6. A missile is fired horizontally from a height of 40,000 ft, with a velocity of 20,000 ft/sec. Find the horizontal distance from the point of firing of the point where the missile hits the ground.

7. Show that the relative velocity of two particles, moving in any directions under the acceleration of gravity, is constant.

N.B. *The velocity of a body A, relative to another body B, is the velocity which A appears to have to an observer moving with B. Thus if* $\mathbf{v}_A$, $\mathbf{v}_B$ *are the velocities of A, B respectively, the velocity of A relative to B is* $\mathbf{v}_A - \mathbf{v}_B$, *while the velocity of B relative to A is* $\mathbf{v}_B - \mathbf{v}_A$.

8. A particle is projected from a point A in the direction $\overrightarrow{AB}$ with the speed u, while a second particle is simultaneously projected from B in the direction $\overrightarrow{BA}$ with speed v. Show that the particles will collide and that the point of collision is vertically below C, the point in AB such that $AC : CB = u : v$.

9. A ball just clears two walls of the same height h and at distances d_1, d_2 from the point of projection. Prove that if α is the angle of projection,

$$\tan \alpha = h(d_1 + d_2)/d_1 d_2.$$

10. A stone is projected horizontally from the top of a tower 80 ft high, with a velocity of 40 ft/sec. At the same moment another stone is projected from the foot of the tower with a velocity of 80 ft/sec in a direction inclined to the horizontal at an angle $\pi/3$. The two stones move in the same vertical plane. Show that the stones collide, and find the position vector of their point of collision.

CHAPTER V

§5.1

If a particle is acted upon by a force directed towards (or away from) a fixed point O, the force is called a *central force*, and the fixed point O is called the *centre of force*. The path of the particle is called its *orbit*.

§5.2

Suppose that at time t, a particle of mass m is at the point P, whose position vector referred to a fixed point O is $\mathbf{r}$. Suppose that the particle is acted upon by a force $\mathbf{F}$ in the direction $\overline{OP}$. Then if $\mathbf{H}$ is the angular momentum of the particle about O,

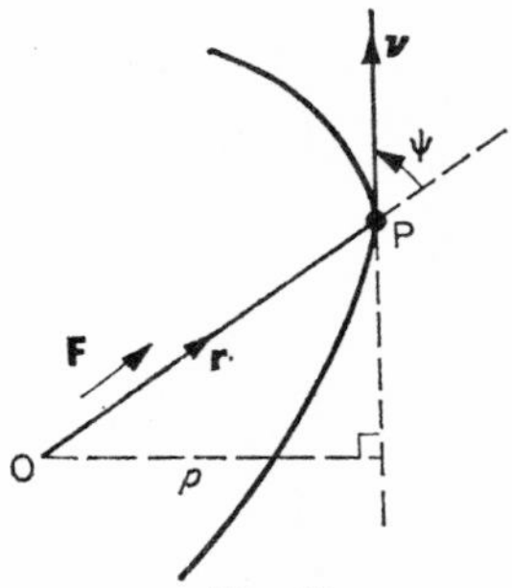

Fig. 29

$$\mathbf{H} = \mathbf{r} \times (m\dot{\mathbf{r}})$$

therefore
$$\frac{d\mathbf{H}}{dt} = \dot{\mathbf{r}} \times (m\dot{\mathbf{r}}) + \mathbf{r} \times (m\ddot{\mathbf{r}})$$
$$= 0 + \mathbf{r} \times \mathbf{F},$$
$$= 0, \text{ since } \mathbf{r}, \mathbf{F} \text{ are in the same direction,}$$

i.e. $\mathbf{H}$ is constant.

Since **H** is a constant vector, its direction is constant and since also $\mathbf{H} = \mathbf{r} \times m\dot{\mathbf{r}}$, **H** is perpendicular to **r**. Hence the vector **r** passes through the fixed point O and is always perpendicular to a fixed direction; i.e. **r** lies in a fixed plane, and the path of the particle is a plane curve. Thus the orbit of a particle moving under the action of a central force is a plane curve.

Suppose $\mathbf{H} = m\mathbf{h}$, where **h** is the angular momentum per unit mass of the particle about O. Then since **H** is a constant vector, and m a constant scalar, **h** also is a constant vector.

Since $$\mathbf{H} = \mathbf{r} \times m\dot{\mathbf{r}},$$

therefore $$m\mathbf{h} = \mathbf{r} \times m\dot{\mathbf{r}},$$

therefore $$\mathbf{h} = \mathbf{r} \times \dot{\mathbf{r}}. \qquad \text{(I)}$$

Using the notation of Chapter IV, equation (I) may be written

$$\mathbf{h} = r\mathbf{l}_1 \times (\dot{r}\mathbf{l}_1 + r\dot{\theta}\mathbf{l}_2)$$

i.e. $$\mathbf{h} = r^2\dot{\theta}\mathbf{l}_3, \qquad \text{(II)}$$

since $$\mathbf{l}_1 \times \mathbf{l}_1 = 0 \quad \text{and} \quad \mathbf{l}_1 \times \mathbf{l}_2 = \mathbf{l}_3.$$

But **h** is a constant vector, hence $r^2\dot{\theta}$ is constant throughout the motion, and $\mathbf{l}_3$ is in a constant direction, normal to the plane of the orbit of the particle.

Suppose that at time t, when the particle is at P, the perpendicular from O to the tangent to the orbit at P is of length p. Then $\mathbf{h} = pv\mathbf{l}_3$ where **v** is the velocity of the particle at P.

Hence

$$h = pv = r^2\dot{\theta} = \text{constant}. \qquad \text{(III)}$$

§5.3

Let **A** be the vector area swept out in time t by the radius OP. Then if $\delta\mathbf{A}$, $\delta\mathbf{r}$, are the small increments in **A**, **r** corresponding to a small rotation $\delta\theta$ of OP, in the small interval of time δt,

$$\delta\mathbf{A} = (\tfrac{1}{2}r^2\delta\theta)\mathbf{l}_3$$

therefore $$\frac{\delta\mathbf{A}}{\delta t} = \tfrac{1}{2}r^2\frac{\delta\theta}{\delta t}\mathbf{l}_3.$$

In the limit as $\delta t \to 0$,

$$\frac{d\mathbf{A}}{dt} = \tfrac{1}{2}r^2\dot{\theta}\mathbf{l}_3 = \tfrac{1}{2}\mathbf{h},$$

i.e.
$$\left|\frac{d\mathbf{A}}{dt}\right| = \tfrac{1}{2}h, \text{ which is constant.} \qquad \text{(IV)}$$

This fact was observed by Kepler in his astronomical observations of the movements of the planets, before it had been demonstrated mathematically, and was stated by Kepler as his second law of planetary motion, in the form: "*The areas described by radii drawn from the sun to a planet are proportional to the times of describing them*", or, more shortly, *the radius joining the sun* (proved to be a centre of force) *to a planet describes equal areas in equal times.*

§5.4. The Inverse Square Law (*First Method*)

Suppose that at any time t a particle is at the point P, whose position vector referred to the fixed point O is $\mathbf{r}$. Suppose that the

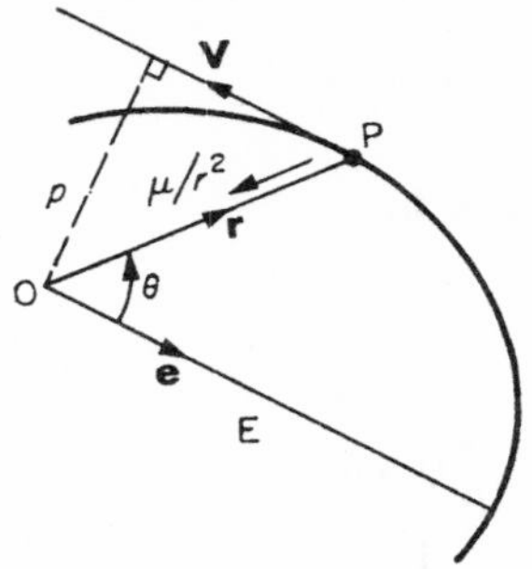

Fig. 30

particle is attracted towards O by a force of magnitude μ/r^2 per unit mass of the particle, where μ is a constant scalar. Then if m is the mass of the particle,

$$m\ddot{\mathbf{r}} = -\frac{\mu m}{r^2}\mathbf{l}_1, \qquad (1)$$

i.e.
$$\frac{1}{\mu}\ddot{\mathbf{r}} = -\frac{1}{r^2}\mathbf{l}_1 \qquad (2)$$

Hence if $\mathbf{h}$ is the constant angular momentum per unit mass of the particle,

$$\frac{1}{\mu}\ddot{\mathbf{r}} \times \mathbf{h} = -\frac{1}{r^2}\mathbf{l}_1 \times \mathbf{h},$$

i.e.
$$\frac{1}{\mu}\ddot{\mathbf{r}} \times \mathbf{h} = -\frac{1}{r^2}\mathbf{l}_1 \times r^2\dot{\theta}\mathbf{l}_3$$

where $\dot{\theta}$ is the angular velocity of OP,

i.e.
$$\frac{1}{\mu}\ddot{\mathbf{r}} \times \mathbf{h} = \dot{\theta}\mathbf{l}_2,$$

$$= \frac{d\mathbf{l}_1}{dt}.$$

Integrating with respect to t, this becomes

$$\frac{1}{\mu}\dot{\mathbf{r}} \times \mathbf{h} = \mathbf{l}_1 + \mathbf{e}, \tag{3}$$

where $\mathbf{e}$ is a constant vector.

From (3),

$$\mathbf{r} \cdot \left\{\frac{1}{\mu}\dot{\mathbf{r}} \times \mathbf{h}\right\} = \mathbf{r} \cdot \{\mathbf{l}_1 + \mathbf{e}\}$$

i.e.
$$\frac{1}{\mu}\{\mathbf{r} \cdot (\dot{\mathbf{r}} \times \mathbf{h})\} = \mathbf{r} \cdot \mathbf{l}_1 + \mathbf{r} \cdot \mathbf{e}$$

i.e.
$$\frac{1}{\mu}\{(\mathbf{r} \times \dot{\mathbf{r}}) \cdot \mathbf{h}\} = \mathbf{r} \cdot \mathbf{l}_1 + \mathbf{r} \cdot \mathbf{e}$$

i.e.
$$\frac{1}{\mu}(\mathbf{h} \cdot \mathbf{h}) = \mathbf{r} \cdot \mathbf{l}_1 + \mathbf{r} \cdot \mathbf{e}$$

i.e.
$$\frac{1}{\mu}h^2 = r + re\cos\theta, \tag{4}$$

where θ is the angle between $\overline{OP}$ and the constant vector $\mathbf{e}$.

Equation (4) may be written.

$$\frac{h^2/\mu}{r} = 1 + e\cos\theta, \tag{5}$$

which is the polar equation of a conic having O as one focus, e as eccentricity, and its major axis in the direction of $\mathbf{e}$.

From equation (5) the semi-latus rectum is h^2/μ. The conic is an ellipse, parabola, or hyperbola according as $e < 1$, $e = 1$, or $e > 1$.

Example 1. *A particle is attracted towards a fixed point O by a force μ/OP^2 per unit mass, where P is the position of the particle at time t. The particle is projected from the point A, where $OA = a$, with velocity $(\mu/2a)^{\frac{1}{2}}$ in a direction making an angle $\pi/4$ with $\overline{OA}$. Find the eccentricity of the orbit and the periodic time of the particle in its orbit.*

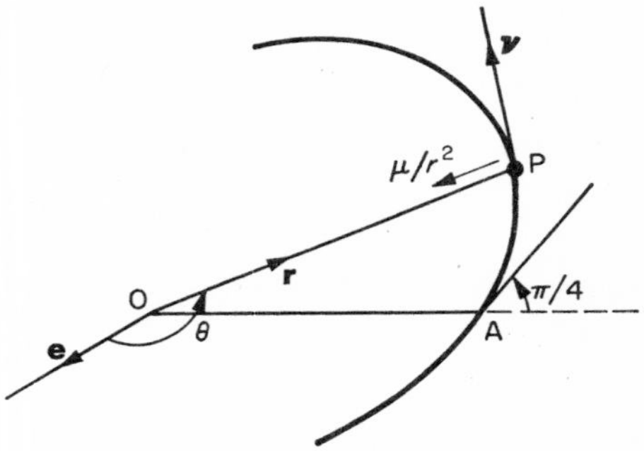

Fig. 31

Let **r** be the position vector of the particle, referred to O, at any time t, and let m be its mass. Let $\mathbf{l}_1$ be the unit vector in the direction $\overline{OP}$, and let $\mathbf{l}_2$ be the unit transverse vector. Let $\mathbf{l}_3$ be the unit vector perpendicular to $\mathbf{l}_1$, $\mathbf{l}_2$ so that $\mathbf{l}_1$, $\mathbf{l}_2$, $\mathbf{l}_3$ form a right-handed system of mutually perpendicular unit vectors. Let **h** be the angular momentum per unit mass of the particle about O. Then **h** is a constant vector such that

$$\mathbf{h} = r^2\dot{\theta}\mathbf{l}_3 = \left(a \sin \frac{\pi}{4}\right)\left(\frac{\mu}{2a}\right)^{\frac{1}{2}} \mathbf{l}_3$$

where $\dot{\theta}$ is the angular velocity of OP,

i.e. $$\mathbf{h} = r^2\dot{\theta}\mathbf{l}_3 = \tfrac{1}{2}(a\mu)^{\frac{1}{2}}\mathbf{l}_3 \qquad (1)$$

Considering the force acting on the particle,

$$m\ddot{\mathbf{r}} = -\frac{\mu}{r^2} m\,\mathbf{l}_1, \tag{2}$$

i.e.

$$\frac{1}{\mu}\ddot{\mathbf{r}} = -\frac{1}{r^2}\mathbf{l}_1. \tag{3}$$

Therefore

$$\frac{1}{\mu}\ddot{\mathbf{r}} \times \mathbf{h} = -\frac{1}{r^2}\mathbf{l}_1 \times \mathbf{h}$$

i.e.

$$\frac{1}{\mu}\ddot{\mathbf{r}} \times \mathbf{h} = -\frac{1}{r^2}\mathbf{l}_1 \times r^2\dot{\theta}\,\mathbf{l}_3,$$

$$= -\dot{\theta}\,\mathbf{l}_1 \times \mathbf{l}_3,$$

$$= \dot{\theta}\,\mathbf{l}_2$$

therefore

$$\frac{1}{\mu}\ddot{\mathbf{r}} \times \mathbf{h} = \frac{d\mathbf{l}_1}{dt}. \tag{4}$$

Integrating with respect to t, equation (4) becomes

$$\frac{1}{\mu}\dot{\mathbf{r}} \times \mathbf{h} = \mathbf{l}_1 + \mathbf{e}, \tag{5}$$

where $\mathbf{e}$ is a constant vector.

Let $\mathbf{l}_{11}$, $\mathbf{l}_{21}$, $\mathbf{l}_{31}$ be the initial values of the unit vectors $\mathbf{l}_1$, $\mathbf{l}_2$, $\mathbf{l}_3$; then if $\dot{\mathbf{r}}_1$ is the initial value of $\dot{\mathbf{r}}$,

$$\dot{\mathbf{r}}_1 = \left(\frac{\mu}{2a}\right)^{\frac{1}{2}} \left(\cos\frac{\pi}{4}\right) \mathbf{l}_{11} + \left(\frac{\mu}{2a}\right)^{\frac{1}{2}} \left(\sin\frac{\pi}{4}\right) \mathbf{l}_{21},$$

i.e.

$$\dot{\mathbf{r}}_1 = \frac{1}{2}\left(\frac{\mu}{a}\right)^{\frac{1}{2}} (\mathbf{l}_{11} + \mathbf{l}_{21}).$$

Substituting in equation (5)

$$\left(\frac{1}{\mu}\right)\left(\frac{1}{2}\right)\left(\frac{\mu}{a}\right)^{\frac{1}{2}} (\mathbf{l}_{11} + \mathbf{l}_{21}) \times (\tfrac{1}{2})\,(a\mu)^{\frac{1}{2}}\,\mathbf{l}_{31} = \mathbf{l}_{11} + \mathbf{e}$$

i.e.

$$\mathbf{e} = -\tfrac{1}{4}\mathbf{l}_{21} + \tfrac{1}{4}\mathbf{l}_{11} - \mathbf{l}_{11}$$

i.e.

$$\mathbf{e} = -\tfrac{3}{4}\mathbf{l}_{11} - \tfrac{1}{4}\mathbf{l}_{21}, \tag{6}$$

and

$$e = \frac{\sqrt{10}}{4} \tag{7}$$

From equation (5)

$$\mathbf{r}\,.\left(\frac{1}{\mu}\dot{\mathbf{r}}\times\mathbf{h}\right)=\mathbf{r}\,.\,\mathbf{l}_1+\mathbf{r}\,.\,\mathbf{e}$$

i.e.
$$\frac{1}{\mu}\{\mathbf{r}\,.\,(\dot{\mathbf{r}}\times\mathbf{h})\}=\mathbf{r}\,.\,\mathbf{l}_1+\mathbf{r}\,.\,\mathbf{e}$$

i.e.
$$\frac{1}{\mu}\{(\mathbf{r}\times\dot{\mathbf{r}})\,.\,\mathbf{h}\}=\mathbf{r}\,.\,\mathbf{l}_1+\mathbf{r}\,.\,\mathbf{e}$$

i.e.
$$\frac{1}{\mu}(\mathbf{h}\,.\,\mathbf{h})=\mathbf{r}\,.\,\mathbf{l}_1+\mathbf{r}_2\,\mathbf{e},$$

i.e.
$$\frac{h^2}{\mu}=r+re\cos\theta$$

where θ is the angle between **r** and the constant vector **e**,

i.e.
$$\frac{h^2}{\mu}=r(1+e\cos\theta) \qquad (8)$$

which is the polar equation of a conic having O as one focus, e as eccentricity, and h^2/μ as semi-latus rectum, the major axis of the conic being in the direction of the vector **e**. Since $e<1$, $(e=(\sqrt{10})/4)$, equation (8) represents an ellipse of eccentricity $(\sqrt{10})/4$.

If α, β, are the semi-major and semi-minor axes of this ellipse, then

$$\frac{h^2}{\mu}=\alpha(1-e^2)$$

i.e.
$$\frac{a\mu}{4\mu}=\alpha\left(1-\frac{10}{16}\right)$$

i.e.
$$\alpha=\frac{2a}{3}$$

and
$$\beta^2=\frac{4a^2}{9}\left(1-\frac{10}{16}\right)$$

i.e.
$$\beta=\frac{a\sqrt{6}}{6}.$$

Since the rate at which OP describes areas is constant, and equal to $\frac{1}{2}h$, i.e. $\frac{1}{4}\sqrt{(a\mu)}$, and the area of the ellipse is $\pi\alpha\beta$, then the time

T required by the particle to travel completely round its orbit is given by

$$T = \frac{\pi\alpha\beta}{h/2},$$

$$= \pi\left(\frac{2a}{3}\right)\left(\frac{a\sqrt{6}}{6}\right)\left(\frac{4}{\sqrt{(a\mu)}}\right),$$

i.e.

$$T = \frac{4\pi}{9}\sqrt{\left(\frac{6a^3}{\mu}\right)}.$$

§5.5. The Inverse Square Law (*Second Method*)

Equation (1) of §5.4 may be written

$$\ddot{\mathbf{r}} = -\frac{\mu}{r^2}\mathbf{l}_1. \qquad (1)$$

Hence

$$2\dot{\mathbf{r}}\,.\,\ddot{\mathbf{r}} = -\left(\frac{\mu}{r^2}\right)2\dot{\mathbf{r}}\,.\,\mathbf{l}_1,$$

i.e.

$$2\dot{\mathbf{r}}\,.\,\ddot{\mathbf{r}} = -\frac{2\mu}{r^2}(\dot{r}\mathbf{l}_1 + r\dot{\theta}\mathbf{l}_2)\,.\,\mathbf{l}_1,$$

i.e.

$$2\dot{\mathbf{r}}\,.\,\ddot{\mathbf{r}} = -\frac{2\mu}{r^2}\dot{r}. \qquad (2)$$

Integrating with respect to t, equation (2) becomes

$$\dot{\mathbf{r}}\,.\,\dot{\mathbf{r}} = \frac{2\mu}{r} + c, \qquad (3)$$

where c is a scalar constant. If $\mathbf{v}$ $(= \dot{\mathbf{r}})$ is the velocity of the particle at time t, equation (3) may be written.

$$v^2 = \frac{2\mu}{r} + c,$$

i.e.

$$\frac{h^2}{p^2} = \frac{2\mu}{r} + c,$$

i.e.

$$\frac{h^2/c}{p^2} = \frac{2\mu/c}{r} + 1,$$

which is the (p, r) equation of a conic.

Example 2. Applying this method to the example in §5.4, equation (1) of §5.4 gives

$$\ddot{\mathbf{r}} = -\frac{\mu}{r^2}\mathbf{l}_1, \tag{1}$$

therefore $$2\dot{\mathbf{r}} \,.\, \ddot{\mathbf{r}} = -\frac{\mu}{r^2} 2\dot{\mathbf{r}} \,.\, \mathbf{l}_1$$

i.e. $$2\dot{\mathbf{r}} \,.\, \ddot{\mathbf{r}} = -\frac{2\mu}{r^2}\dot{r}. \tag{2}$$

Integrating with respect to t, equation (2) becomes

$$\dot{\mathbf{r}} \,.\, \dot{\mathbf{r}} = \frac{2\mu}{r} + c,$$

where c is a scalar constant,

i.e. $$v^2 = \frac{2\mu}{r} + c, \tag{3}$$

where $$\mathbf{v} = \dot{\mathbf{r}}.$$

But initially $v = \sqrt{\left(\frac{\mu}{2a}\right)}$ and $r = a$,

therefore $$\frac{\mu}{2a} = \frac{2\mu}{a} + c,$$

therefore $$c = -\frac{3\mu}{2a}.$$

Hence equation (3) may be written

$$v^2 = \frac{2\mu}{r} - \frac{3\mu}{2a}. \tag{4}$$

Since $pv = h = \frac{1}{2}\sqrt{(a\mu)}$, equation (4) may be written

$$\frac{a\mu}{4p^2} = \frac{2\mu}{r} - \frac{3\mu}{2a}$$

i.e. $$\frac{a^2/6}{p^2} = \frac{4a/3}{r} - 1$$

which is the (p, r) equation of an ellipse whose semi-major axis, α, and semi-minor axis, β, are given by

$$\alpha = \frac{2a}{3}, \qquad \beta = \frac{a}{\sqrt{6}}$$

Hence the eccentricity e is given by

$$\beta^2 = \alpha^2(1 - e^2);$$

i.e. $$\frac{a^2}{6} = \frac{4a^2}{9}(1 - e^2);$$

i.e. $$1 - e^2 = \frac{3}{8};$$

therefore $$e^2 = \frac{5}{8};$$

and $$e = \frac{\sqrt{10}}{4}.$$

The periodic time, T, is obtained from the relation

$$T = \frac{2\pi\alpha\beta}{h}$$

as in §5.4.

§5.6. Laws of Force Other than the Inverse Square Law

An examination of the methods demonstrated in §5.4 and §5.5, shows that the method of §5.4 can only conveniently be used when the law of force is the inverse square law, while the method of §5.5 has a wider general application, but has the disadvantage of giving the equation of the orbit in the (p, r) form which is less familiar than the polar equation given by the method of §5.4.

An example should suffice to illustrate the method of §5.5 when the law of force is not the inverse square law.

Example 3. *A particle is attracted towards a fixed point O by a force* $(\mu/r^2) + (\mu a/4r^3)$ *per unit mass, where r is the distance of the particle from O at any time t. The particle is projected from the point A, whose distance from O is a, with velocity* $\sqrt{(\mu/3a)}$ *per-*

pendicular to OA. Find the (p, r) equation of the orbit of the particle, and show that its distance from O never exceeds a and is never less than $a/23$.

Suppose that the particle is at the point P at any time t, and that the position vector of P referred to O is $\mathbf{r}$. Then with the usual notation,

$$\ddot{\mathbf{r}} = -\mu\left(\frac{1}{r^2} + \frac{a}{4r^3}\right)\mathbf{l}_1, \tag{1}$$

therefore $$2\dot{\mathbf{r}} \,.\, \ddot{\mathbf{r}} = -2\mu\left(\frac{1}{r^2} + \frac{a}{4r^3}\right)\dot{\mathbf{r}} \,.\, \mathbf{l}_1$$

i.e. $$2\dot{\mathbf{r}} \,.\, \ddot{\mathbf{r}} = -2\mu\left(\frac{1}{r^2} + \frac{a}{4r^3}\right)(\dot{r}\mathbf{l}_1 + r\dot{\theta}\mathbf{l}_2) \,.\, \mathbf{l}_1,$$

where $\dot{\theta}$ is the angular velocity of OP,

i.e. $$2\dot{\mathbf{r}} \,.\, \ddot{\mathbf{r}} = -2\mu\left(\frac{1}{r^2} + \frac{a}{4r^3}\right)\dot{r}. \tag{2}$$

Integrating, equation (2) becomes

$$\dot{\mathbf{r}} \,.\, \dot{\mathbf{r}} = \frac{2\mu}{r} + \frac{\mu a}{4r^2} + c,$$

where c is a constant scalar. Then if $\mathbf{v} = \dot{\mathbf{r}}$,

$$v^2 = \frac{2\mu}{r} + \frac{\mu a}{4r^2} + c. \tag{3}$$

But $v = \sqrt{(\mu/3a)}$ when $r = a$

therefore $$c = \frac{\mu}{3a} - \frac{2\mu}{a} - \frac{\mu}{4a},$$

i.e. $$c = -\frac{23\mu}{12a}.$$

Then equation (3) may be written

$$v^2 = \frac{2\mu}{r} + \frac{\mu a}{4r^2} - \frac{23\mu}{12a}. \tag{4}$$

If $\mathbf{h}$ is the angular momentum per unit mass of the particle about O, and p is the length of the perpendicular from O to the tangent at P to the path of the particle, then

$$pv = h = a\sqrt{\left(\frac{\mu}{3a}\right)}$$

Hence equation (4) may be written

$$\frac{\mu a}{3p^2} = \frac{2\mu}{r} + \frac{\mu a}{4r^2} - \frac{23\mu}{12a}$$

i.e. $$\frac{a}{3p^2} = \frac{2}{r} + \frac{a}{4r^2} - \frac{23}{12a} \qquad (5)$$

From equation (5) since $p \leqslant r$,

therefore $$\frac{a}{3r^2} \leqslant \frac{2}{r} + \frac{a}{4r^2} - \frac{23}{12a}$$

i.e. $$4a^2 \leqslant 24ar + 3a^2 - 23r^2$$

i.e. $$a^2 - 24ar + 23r^2 \leqslant 0$$

i.e. $$(a - r)(a - 23r) \leqslant 0$$

i.e. $$\frac{a}{23} \leqslant r \leqslant a.$$

Hence, the distance r of the particle from O never exceeds a and is never less than $a/23$.

Examples V

1. A particle of mass m is attracted towards the fixed point S by a force $3mcu^2/r^2$ where r is the distance of the particle from S at any time t, and c, u are constants. Initially the particle is projected from a point C with velocity u in a direction inclined to $\overline{SC}$ at an angle $\pi/3$, the distance SC being c. Find the eccentricity of the orbit of the particle and determine whether it is an ellipse parabola, or hyperbola.

2. A particle is attracted towards a fixed point O by a force μ/OP^2 per unit mass, where P is the position of the particle at time t. Write down the vector equation of motion of the particle, and prove that its path is a conic with one focus at O.

If the particle is projected from the point A, where $OA = a$, with velocity $\sqrt{(\mu/a)}$ in a direction making an angle $\pi/4$ with $\overline{OA}$, find the eccentricity of the orbit, and the periodic time of the particle in its orbit.

3. A particle moves under the action of a central force attracting the particle towards a point O with a force inversely proportional to its distance from O, the force of attraction being 32 pdl per unit mass when the distance is 2 ft. If the particle is projected from a point A in a direction perpendicular to OA with a velocity of 16 ft/sec, OA being 1 ft, determine the equation of the orbit.

4. A particle of mass M moves under the action of an attractive central force $\mu M/r^2$. Show that the path of the particle is a conic section whose semi-latus rectum is h^2/μ, where $M\mathbf{h}$ is the angular momentum of the particle about the centre of attraction.

The particle originally moves in a circular orbit. Show that if μ is instantaneously decreased by a fraction c of its original value, then in the subsequent motion the ratio of the distance of the particle from the centre of force at the perihelion to its distance at the aphelion is $(1 - 2c) : 1$.

N.B. *In planetary motion the perihelion is the point of the orbit which is nearest to the centre of attraction and the aphelion is the point of the orbit which is furthest from the centre of attraction.*

5. A particle P is projected, with speed V, from a point at a distance R from a centre of force O, and is attracted towards O by a force μ/OP^2 per unit mass. Show that the path of the particle is an ellipse, parabola, or hyperbola according as V^2 is less than, equal to, or greater than $2\mu/R$.

Two particles of masses M and m, describing coplanar parabolic orbits about a centre of force at their common focus, impinge at right angles, and coalesce, at a distance R from the centre of force. Show that the path of the composite particle is an ellipse whose major axis is of length $R(M + m)^2/2Mm$.

6. A particle P of unit mass is attracted towards a fixed point S by a force of magnitude μ/SP^2. The particle is projected from a point P_o with speed u (*not* in the line SP_o). Prove that the path of the particle is a conic with S as focus.

Find the conditions that the path should be (i) elliptic, (ii) hyperbolic, (iii) parabolic.

Show that if the path is elliptic, the periodic time is

$$2\pi\mu\left\{\frac{SP_o}{2\mu - u^2SP_o}\right\}^{3/2}$$

7. Obtain expressions for the radial and transverse components of velocity and acceleration of a particle P moving in a plane.

P has velocities u, v of constant magnitudes, u in a fixed direction and v perpendicular to OP where O is a fixed point. Show that the acceleration of the particle is always directed towards O, and varies inversely as the square of the distance from O. Find the eccentricity of the orbit in terms of u, v.

8. A particle describes an ellipse under an attraction μr per unit mass towards a fixed point O. Prove that, with the usual notation,

$$v^2 = \mu(a^2 + b^2 - r^2), \quad h = ab\sqrt{\mu}$$

An equal particle, subject to the same attractive force, is projected from O with speed $b\sqrt{(11\mu)}$ along the major axis of the ellipse. The two particles collide and coalesce at the end of the major axis. If $a = 2b$ prove that the orbit of the composite particle is an ellipse whose axes are $2b(\sqrt{2} \pm 1)$.

9. A missile P is projected from a point A on the earth's surface with speed $\sqrt{(2gd)}$, the earth being supposed spherical, of radius $c(c > d)$, and centre O. The missile is assumed to be attracted towards O by a force gc^2/r^2 per unit mass where $OP = r$. Show that the path of P is an ellipse, having O as focus, and that the speed v of P is given by

$$v^2 = 2g\left(d - c + \frac{c^2}{r}\right).$$

Show that the locus of the second focus S, for all paths in which the particle leaves A in the same vertical plane with the same speed, is a circle of centre A and radius $cd/(c - d)$.

10. A particle P describes an ellipse under an attraction towards a focus S, Prove that the attraction varies inversely as SP^2. Prove also that the Kinetic Energy of the particle is proportioned to $(2/SP) - (1/a)$ where a is the semi-major axis of the ellipse. If the greatest and least speeds of the particle are v_1 and v_2 respectively, show that the eccentricity of the path is $(v_1 - v_2)/(v_1 + v_2)$.

11. A particle P moves in the plane of the rectangular axes OX, OY under an attraction n^2OP per unit mass towards O. When the particle is at the point $(a, 0)$ its velocity is u in the direction OY. Show that the path of the particle is an ellipse.

Find the velocity of the particle at time t after leaving $(a, 0)$, and also the periodic time of the particle in its orbit.

12. A particle P is moving under a central force μ/r^2 per unit mass towards a fixed point S where $SP = r$. Initially the particle is at a distance R from S and has a velocity V in a direction making an angle ϕ with SP. If $V^2R = 2a\mu$ show that the path of P is a conic with one focus at S, semi-latus rectum $2aR \sin^2 \phi$ and eccentricity e where $e^2 = 1 - 4a(1 - a) \sin^2 \phi$.

13. A particle is attracted to a fixed point O by a force varying inversely as the square of the distance of the particle from O. Prove that the orbit is a conic with one focus at O and, with the usual notation, semi-latus rectum h^2/μ. Show also that the speed v is given by

$$v^2 = \frac{2\mu}{r} + \frac{\mu(e^2 - 1)}{l}$$

where l is the semi-latus rectum.

While describing a circular orbit with speed U, the particle explodes into two equal parts. The initial velocity of separation of the two parts is in the line of the velocity immediately before the explosion and is of magnitude $2V$. Show that if $V < (\sqrt{2} - 1)U$, both parts subsequently describe elliptic orbits.

14. (*a*) If $\mathbf{r}$ is any vector function of t, prove that

(i) $$r\frac{dr}{dt} = \mathbf{r}\,.\,\dot{\mathbf{r}};$$

(ii) $$r^3\frac{d}{dt}\left(\frac{1}{r}\right) = -\,\mathbf{r}\;\dot{\mathbf{r}};$$

(iii) $$r^3\frac{d}{dt}\left(\frac{\mathbf{r}}{r}\right) = -\,\mathbf{r} \times (\mathbf{r} \times \dot{\mathbf{r}}).$$

(*b*) The sun attracts a planet of mass m with a force of magnitude MmG/r^2, where M is the mass of the sun, G is a constant, and r the distance of the planet from the sun. Write down the equation of motion of the planet relative to the sun, which may be assumed to be fixed, and show that the speed v of the planet is given by

$$\frac{1}{2}mv^2 - \frac{mMG}{r} = E,$$

where E is a constant.

When the planet is at a distance $2a$ from the sun its speed is V, and when half this distance away, its speed is $2V$. Show that the mass of the sun is $3V^2a/G$.

15. A particle of mass M has the position vector $\mathbf{r}$ referred to a fixed point O, and is acted on by a force $-M\mu(r)^{-3}\mathbf{r}$. Initially it is given a velocity $\mathbf{V}$ such that $V^2 < 2\mu/a$, where $\mathbf{a}$ is the position vector of the initial position of the particle. Show that the path of the particle is an ellipse and that the velocity $\mathbf{v}$ of the particle at any time t satisfies the relation

$$\mathbf{h} \times \mathbf{v} = -\frac{\mu}{r}\mathbf{r} + \mu e \mathbf{e}_1,$$

where $M\mathbf{h}$ is the angular momentum about O, e the eccentricity of the ellipse and $\mathbf{e}_1$ the unit vector in the direction of the major axis. Hence by evaluating $\mathbf{h} \times (\mathbf{h} \times \mathbf{v})$, or otherwise, show that $\mathbf{v}$ can be expressed as the sum of a vector of constant magnitude perpendicular to $\mathbf{r}$, and a constant vector perpendicular to $\mathbf{e}_1$.

16. A particle moves in an elliptic orbit of major axis $2a$ under a central attraction μ/r^2 per unit mass. Prove that its velocity $\mathbf{v}$ is given by

$$v^2 = \mu\left(\frac{2}{r} - \frac{1}{a}\right),$$

where the relation $h^2 = \mu l$, using the usual notation, may be assumed.

A satellite is moving uniformly in a circular orbit of radius $4c$ about the earth's centre O, the radius of the earth being c. The speed of the satellite is suddenly reduced in magnitude, its direction remaining unaltered, so that it subsequently pursues an elliptic orbit such that its least distance from O in the subsequent motion is $2c$. Show that the speed must be reduced in the ratio $\sqrt{2} : \sqrt{3}$. Show also that the time taken by the satellite to reach this distance $2c$ from O is $\pi\sqrt{(27c/g)}$ and find its speed at this point in terms of g and c, g being the acceleration due to gravity at the earth's surface.

CHAPTER VI

§6.1

It will be convenient to use the notation "the point A (**a**)" to denote "the point A whose position vector referred to the origin O is **a**".

§6.2. Equation of a Straight Line Through a Given Point in a Given Direction

Let O be the origin, and let P (**r**) be any point on the line AB drawn through A (**a**), in the direction of the vector **b**. Then there is a number t, dependent on the position of P, such that $\overline{AP} = t\mathbf{b}$.

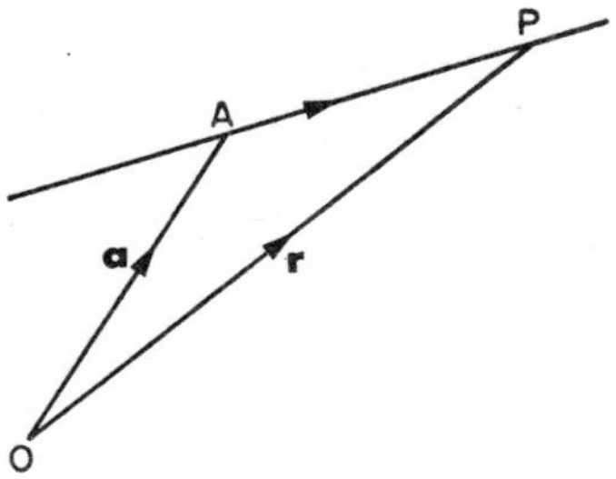

Fig. 32

Then $$\overline{OP} = \overline{OA} + \overline{AP},$$

i.e. $$\mathbf{r} = \mathbf{a} + t\mathbf{b}. \quad \text{(I)}$$

For any point P on the line AB there is a unique value of t; also for any value of t there is a unique point P on the line AB.

For suppose t_1 is a particular value of t, then the point P_1 determined by equation (I) has position vector $\mathbf{r}_1$ where

$$\mathbf{r}_1 = \mathbf{a} + t_1\mathbf{b},$$

i.e. $$\overline{OP_1} = \overline{OA} + t_1\mathbf{b}$$

i.e. $$\overline{OP_1} - \overline{OA} = t_1\mathbf{b},$$

i.e. $$\overline{AP_1} = t_1\mathbf{b}.$$

Hence the line AP_1 passes through A and is parallel to $\mathbf{b}$; i.e. the line AP_1 coincides with AB, and therefore P_1 lies on AB. Thus any point P, such that $\overline{AP} = t\mathbf{b}$, whose parameter t, satisfies equation (I) lies on AB. Hence equation (I) is the equation of the line through A in the direction of $\mathbf{b}$.

§6.3. Equation of a Straight Line through Two Given Points

Let O be the origin and let A ($\mathbf{a}$), B ($\mathbf{b}$) be the two given points. Let P ($\mathbf{r}$) be any point on the line AB. Then for any position of P on AB there must be a number t such that $AP : AB = t : 1$.

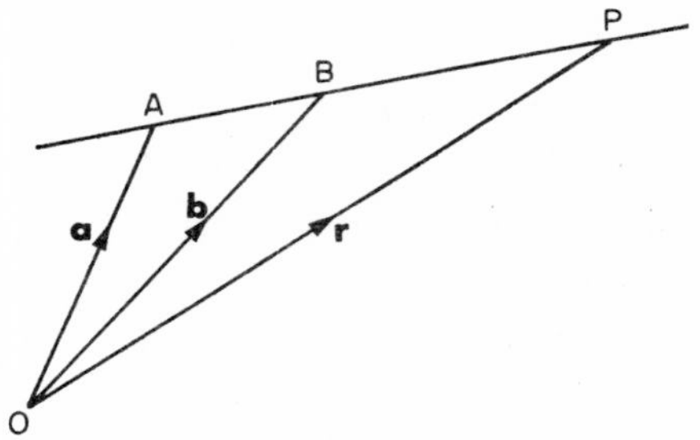

Fig. 33

therefore $$\overline{AP} = t\overline{AB},$$

i.e. $$\mathbf{r} - \mathbf{a} = t(\mathbf{b} - \mathbf{a})$$

i.e. $$\mathbf{r} = \mathbf{a} + t(\mathbf{b} - \mathbf{a}) \qquad \text{(II)}$$

or $$\mathbf{r} = (1 - t)\mathbf{a} + t\mathbf{b}. \qquad \text{(III)}$$

Hence the parameter t of any point P on AB, where $AP : AB = t : 1$, satisfies equations (II), (III). By an argument similar to that of §6.2 it can be shown that if P is a point such that $AP : AB = t : 1$, and the parameter t satisfies equation (II), then P must lie on AB.

§6.4. Equation of a Plane Through a Given Point and Parallel to Two Given Vectors

Let O be the origin and let II be the plane parallel to the vectors **a**, **b** containing the given point C (**c**). Then lines CA, CB may be drawn in plane II so that $\overline{CA} = \mathbf{a}$ and $\overline{CB} = \mathbf{b}$.

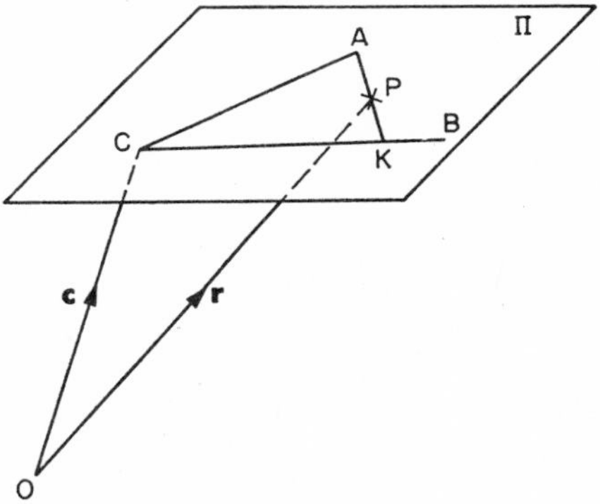

Fig. 34

Let K (**k**) be an arbitrary point in CB such that $\overline{CK} = \lambda\mathbf{b}$. Then $\overline{OA} = \mathbf{c} + \mathbf{a}$ and $\overline{OK} = \mathbf{c} + \lambda\mathbf{b}$. If P (**r**) is any point on the line AK, there must be a number μ such that,

$$\mathbf{r} = (\mathbf{c} + \mathbf{a}) + \mu\{(\mathbf{c} + \lambda\mathbf{b}) - (\mathbf{c} + \mathbf{a})\}$$

from equation (II) of §6.3,

i.e. $$\mathbf{r} = \mathbf{c} + \mathbf{a} + \mu\{\lambda\mathbf{b} - \mathbf{a}\},$$

i.e. $$\mathbf{r} = \mathbf{c} + (1 - \mu)\mathbf{a} + \lambda\mu\mathbf{b}.$$

By a suitable choice of λ, μ, the point P can be made to take any position in plane Π. Hence the equation

$$\mathbf{r} = \mathbf{c} + (1 - \mu)\mathbf{a} + \lambda\mu\mathbf{b}$$

gives the position vector of an arbituary point P in plane Π.

Writing s instead of $(1 - \mu)$ and t instead of $\lambda\mu$, the equation of plane Π can be written in the form

$$\mathbf{r} = \mathbf{c} + s\mathbf{a} + t\mathbf{b}. \tag{IV}$$

By an argument similar to that of §6.2, it can be shown that if a pair of values of s and t are substituted in equation (IV), the point so determined lies in the plane through A parallel to the vectors **a, b.**

§6.5. Equation of a Plane Containing Three Given Points

Let O be the origin, and let A (**a**), B (**b**), C (**c**) be the three given

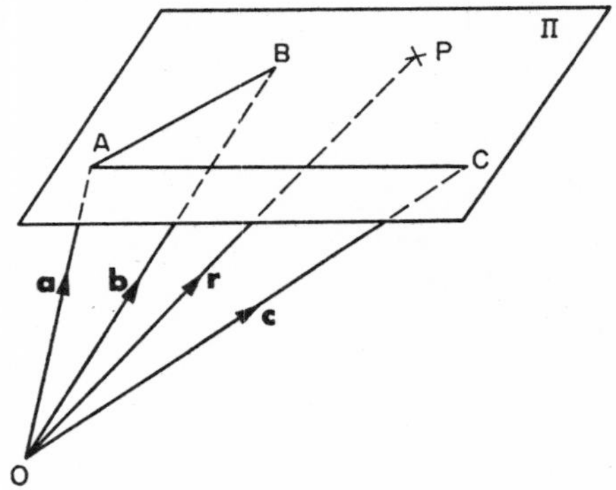

Fig. 35

points which define the plane Π. Then $\overline{AB} = \mathbf{b} - \mathbf{a}$ and $\overline{AC} = \mathbf{c} - \mathbf{a}$. Hence Π is the plane, containing the point A (**a**), parallel to the vectors $(\mathbf{b} - \mathbf{a})$ and $(\mathbf{c} - \mathbf{a})$. Hence (§6.4) the equation of plane Π is

$$\mathbf{r} = \mathbf{a} + s(\mathbf{b} - \mathbf{a}) + t(\mathbf{c} - \mathbf{a}),$$

i.e.

$$\mathbf{r} = (1 - s - t)\mathbf{a} + s\mathbf{b} + t\mathbf{c}, \tag{V}$$

where s, t are parameters.

§6.6. Equation of a Plane Containing a Given Point and Normal to a Given Direction

Let O be the origin and let A ($\mathbf{a}$) be the given point. Let $\mathbf{n}$ be unit vector in the given direction, and let p be the length of the

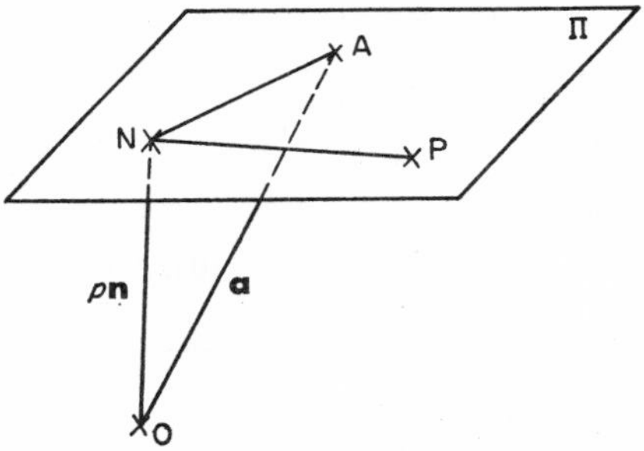

Fig. 36

perpendicular from O to the plane through A normal to $\mathbf{n}$. Let N be the foot of the perpendicular from O to Π. Then the position vector of N is $p\mathbf{n}$. If P ($\mathbf{r}$) is any point in plane Π, $\overline{AP}$ is perpendicular to $\overline{ON}$,

i.e. $$\overline{AP} \cdot \mathbf{n} = 0,$$

i.e. $$(\mathbf{r} - \mathbf{a}) \cdot \mathbf{n} = 0$$

or $$\mathbf{r} \cdot \mathbf{n} = \mathbf{a} \cdot \mathbf{n} \qquad \text{(VI)}$$

whence it can be seen that the length of the perpendicular from the origin to plane Π is $\mathbf{a} \cdot \mathbf{n}$.

§6.7. Alternative Forms of the Equation of a Plane and of its Distance from the Origin

Let A ($\mathbf{a}$), B ($\mathbf{b}$), C ($\mathbf{c}$) be three points in the plane Π. Let P ($\mathbf{r}$) be any point in Π. Then the vectors $\overline{AP}$, $\overline{BA}$, $\overline{CB}$ are coplanar; i.e. the vectors $(\mathbf{r} - \mathbf{a})$, $(\mathbf{a} - \mathbf{b})$, $(\mathbf{b} - \mathbf{c})$ are coplanar.

therefore $$[(\mathbf{r}-\mathbf{a})(\mathbf{a}-\mathbf{b})(\mathbf{b}-\mathbf{c})]=0$$

i.e. $$(\mathbf{r}-\mathbf{a})\,.\,\{(\mathbf{a}-\mathbf{b})\times(\mathbf{b}-\mathbf{c})\}=0,$$

i.e. $$(\mathbf{r}-\mathbf{a})\,.\,\{\mathbf{a}\times\mathbf{b}+\mathbf{c}\times\mathbf{a}+\mathbf{b}\times\mathbf{c}\}=0,$$

i.e. $$\mathbf{r}\,.\,(\mathbf{b}\times\mathbf{c}+\mathbf{c}\times\mathbf{a}+\mathbf{a}\times\mathbf{b})-\mathbf{a}\,.\,(\mathbf{b}\times\mathbf{c})=0,$$

which may be written

$$\mathbf{r}\,.\,(\mathbf{b}\times\mathbf{c}+\mathbf{c}\times\mathbf{a}+\mathbf{a}\times\mathbf{b})=[\mathbf{a}\,\mathbf{b}\,\mathbf{c}], \qquad \text{(VIII)}$$

or $$[\mathbf{r}\,\mathbf{b}\,\mathbf{c}]+[\mathbf{r}\,\mathbf{c}\,\mathbf{a}]+[\mathbf{r}\,\mathbf{a}\,\mathbf{b}]=[\mathbf{a}\,\mathbf{b}\,\mathbf{c}] \qquad \text{(IX)}$$

If **n** is the unit vector normal to plane II, and Δ is the area of triangle ABC, then,

$$\overline{BA}\times\overline{CB}=\pm\,2\Delta\mathbf{n},$$

the positive or negative sign being used according as $\overline{BA}$, $\overline{CB}$, **n**, do, or do not, form a right-handed system,

i.e. $$(\mathbf{a}-\mathbf{b})\times(\mathbf{b}-\mathbf{c})=\pm\,2\Delta\mathbf{n},$$

i.e. $$\mathbf{a}\times\mathbf{b}-\mathbf{a}\times\mathbf{c}+\mathbf{b}\times\mathbf{c}=\pm\,2\Delta\mathbf{n},$$

i.e. $$\mathbf{b}\times\mathbf{c}+\mathbf{c}\times\mathbf{a}+\mathbf{a}\times\mathbf{b}=\pm\,2\Delta\mathbf{n},$$

i.e. $$2\Delta=|\,\mathbf{b}\times\mathbf{c}+\mathbf{c}\times\mathbf{a}+\mathbf{a}\times\mathbf{b}\,| \qquad \text{(X)}$$

Then from §6.6, the length p of the perpendicular from O to plane Π is given by

$$p=\mathbf{a}\,.\,\mathbf{n}$$

$$=\frac{\pm\,\mathbf{a}\,.\,\{\mathbf{b}\times\mathbf{c}+\mathbf{c}\times\mathbf{a}+\mathbf{a}\times\mathbf{b}\}}{2\Delta}$$

i.e. $$p=\pm\,\frac{[\mathbf{a}\,\mathbf{b}\,\mathbf{c}]}{|\,\mathbf{b}\times\mathbf{c}+\mathbf{c}\times\mathbf{a}+\mathbf{a}\times\mathbf{b}\,|} \qquad \text{(XI)}$$

§6.8. Vector Area and Condition for Collinearity

Suppose A (**a**), B (**b**), C (**c**) are any three points. Then $\overline{AB} = \mathbf{b} - \mathbf{a}$, $\overline{AC} = \mathbf{c} - \mathbf{a}$, and if Δ is the area of triangle ABC, then

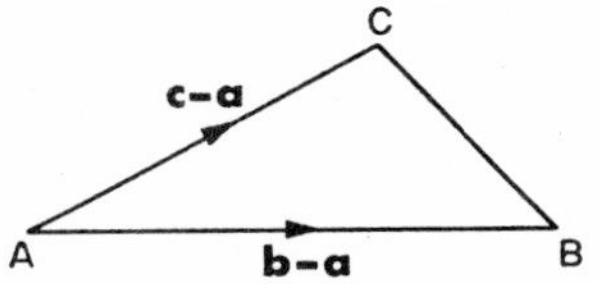

Fig. 37

from §2.14 $\Delta\mathbf{n} = \frac{1}{2}(\mathbf{b} - \mathbf{a}) \times (\mathbf{c} - \mathbf{a})$, where $\mathbf{n}$ is unit vector normal to the plane of triangle ABC, such that $\overline{AB}$, $\overline{AC}$, $\mathbf{n}$, form a right-handed system.

Then $$\Delta\mathbf{n} = \tfrac{1}{2}\{\mathbf{b} \times \mathbf{c} - \mathbf{b} \times \mathbf{a} - \mathbf{a} \times \mathbf{c}\},$$

i.e. $$\Delta\mathbf{n} = \tfrac{1}{2}\{\mathbf{b} \times \mathbf{c} + \mathbf{c} \times \mathbf{a} + \mathbf{a} \times \mathbf{b}\}.$$

If the points A, B, C are collinear, the area of triangle ABC is zero, hence a condition for the collinearity of A, B, C is

$$\mathbf{b} \times \mathbf{c} + \mathbf{c} \times \mathbf{a} + \mathbf{a} \times \mathbf{b} = 0$$

Example 1. *Write down the equation of the straight line joining the points* (1, –3, 1) *and* (0, –3, 2). *Write down the equation of the plane containing the origin and the points* (3, 4, 1) *and* (6, 0, 2). *Find the coordinates of the point in which the line cuts the plane.*

The straight line joining the points (1, –3, 1) and (0, –3, 2) has the equation

$$\mathbf{r} = (\mathbf{i}_1 - 3\mathbf{i}_2 + \mathbf{i}_3) + t(-\mathbf{i}_1 + \mathbf{i}_3)$$

where t is the parameter of any point on the line,

i.e. $$\mathbf{r} = (1 - t)\mathbf{i}_1 - 3\mathbf{i}_2 + (1 + t)\mathbf{i}_3. \qquad (1)$$

The plane containing O and the points (3, 4, 1), (6, 0, 2) has the equation

$$\mathbf{r} = 0 + p(3\mathbf{i}_1 + 4\mathbf{i}_2 + \mathbf{i}_3) + q(6\mathbf{i}_1 + 2\mathbf{i}_3),$$

where p, q are the parameters of any point on the plane,

i.e. $$\mathbf{r} = (3p + 6q)\mathbf{i}_1 + 4p\mathbf{i}_2 + (p + 2q)\mathbf{i}_3 \qquad (2)$$

Suppose line (1) meets plane (2) in the point A (**a**) whose parameters are t_1, p_1, q_1. Then

$$1 - t_1 = 3p_1 + 6q_1 \tag{3}$$

$$-3 = 4p_1, \tag{4}$$

$$1 + t_1 = p_1 + 2q_1. \tag{5}$$

From (3) and (5)

$$3(1 + t_1) = 1 - t_1$$

therefore $$4t_1 = -2,$$

therefore $$t_1 = -\tfrac{1}{2}.$$

Hence A is the point whose position vector is **a** where

$$\mathbf{a} = (1 - t_1)\mathbf{i}_1 - 3\mathbf{i}_2 + (1 + t_1)\mathbf{i}_3,$$
$$= \tfrac{3}{2}\mathbf{i}_1 - 3\mathbf{i}_2 + \tfrac{1}{2}\mathbf{i}_3$$

i.e. A is the point $(\tfrac{3}{2}, -3, \tfrac{1}{2})$.

Example 2. *Find the distance from the point* (7, 8, 5) *to the plane which contains the points* (1, 2, 3), (2, 4, 5), (3, 5, 9).

Let A, B, C be the points (1, 2, 3), (2, 4, 5), (3, 5, 9) respectively and let D be the point (7, 8, 5). Let **a**, **b**, **c**, **d**, be the position vectors of A, B, C, D, and let **n** be the unit vector normal to the plane ABC in the direction from O to the plane.

Then $$\mathbf{a} = \mathbf{i}_1 + 2\mathbf{i}_2 + 3\mathbf{i}_3,$$

$$\mathbf{b} = 2\mathbf{i}_1 + 4\mathbf{i}_2 + 5\mathbf{i}_3$$

$$\mathbf{c} = 3\mathbf{i}_1 + 5\mathbf{i}_2 + 9\mathbf{i}_3,$$

$$\mathbf{d} = 7\mathbf{i}_1 + 8\mathbf{i}_2 + 5\mathbf{i}_3.$$

Let $$\mathbf{n} = n_1\mathbf{i}_1 + n_2\mathbf{i}_2 + n_3\mathbf{i}_3$$

where $$n_1^2 + n_2^2 + n_3^2 = 1.$$

$$\overline{AB} = \mathbf{i}_1 + 2\mathbf{i}_2 + 2\mathbf{i}_3$$

$$\overline{AC} = 2\mathbf{i}_1 + 3\mathbf{i}_2 + 6\mathbf{i}_3.$$

Since **n** is perpendicular to both $\overline{AB}$ and $\overline{AC}$,

therefore
$$n_1 + 2n_2 + 2n_3 = 0$$
$$2n_1 + 3n_2 + 6n_3 = 0$$

therefore
$$\frac{n_1}{6} = \frac{n_2}{-2} = \frac{n_3}{-1} = \frac{1}{\sqrt{41}}. \tag{1}$$

Therefore
$$\mathbf{n} = -\frac{6}{\sqrt{41}}\mathbf{i}_1 + \frac{2}{\sqrt{41}}\mathbf{i}_2 + \frac{1}{\sqrt{41}}\mathbf{i}_3 \tag{2}$$

Then if p_1 is the length of the perpendicular from O to the plane ABC,
$$p_1 = \mathbf{a} \,.\, \mathbf{n},$$
$$= -\frac{6}{\sqrt{41}} + \frac{4}{\sqrt{41}} + \frac{3}{\sqrt{41}},$$

i.e.
$$p_1 = \frac{1}{\sqrt{41}}.$$

If p_2 is the length of the perpendicular from O to the plane through D parallel to the plane ABC,
$$p_2 = \mathbf{d} \,.\, \mathbf{n},$$
$$= -\frac{42}{\sqrt{41}} + \frac{16}{\sqrt{41}} + \frac{5}{\sqrt{41}},$$

i.e.
$$p_2 = -\frac{21}{\sqrt{41}}.$$

Therefore the distance from D to the plane ABC is $22/\sqrt{41}$ units.

Note 1. It will be seen that equation (1) gives the result $n_1 : n_2 : n_3 = 6 : (-2) : (-1)$ or $n_1 : n_2 : n_3 = (-6) : 2 : 1$, while $n_1^2 + n_2^2 + n_3^2 = 1$. The signs in equation (2) have been chosen to make **a . n** positive, thus ensuring that the sense of **n** is from O to the plane ABC.

Note 2. The student should note that the choice of signs in equation (2) produces a negative result for the product **d . n**, thus showing that the perpendicular from O to the plane Π through D, parallel to the plane ABC is in the opposite sense to the perpendicular from O to the plane ABC, and that the perpendicular *from O to* plane Π is in fact in the direction of $-\mathbf{n}$. The negative result $\mathbf{d}\,.\,\mathbf{n}$ indicates, therefore, that plane Π is on the side of O remote from plane ABC.

If p_3 is the perpendicular from O to the plane Π' through $E\,(-1, 3, 2)$, then

$$p_3 = \mathbf{e}\,.\,\mathbf{n}$$

where $\mathbf{e}$ is the position vector of E.

Hence
$$p_3 = \frac{6}{\sqrt{41}} + \frac{6}{\sqrt{41}} - \frac{2}{\sqrt{41}}$$

i.e.
$$p_3 = \frac{10}{\sqrt{41}}$$

showing that Π' is on the same side of O as plane ABC, but further away from O.

The distance of E from plane ABC is therefore $9/\sqrt{41}$ units.

Example 3. *Find the coordinates of the point in which the line joining A (2, 3, 4) and B (1, 4, —3) cuts the (x, y) plane.*

The position vector $\mathbf{r}$ of any point P on AB satisfies the equation

$$\mathbf{r} = (2\mathbf{i}_1 + 3\mathbf{i}_2 + 4\mathbf{i}_3) + t(-\mathbf{i}_1 + \mathbf{i}_2 - 7\mathbf{i}_3)$$

i.e.
$$\mathbf{r} = (2 - t)\mathbf{i}_1 + (3 + t)\mathbf{i}_2 + (4 - 7t)\mathbf{i}_3$$

where t is the parameter of P.

If P lies on the (x, y) plane

$$4 - 7t = 0,$$

i.e.
$$t = \frac{4}{7}.$$

Hence
$$\mathbf{r} = \frac{10}{7}\mathbf{i}_1 + \frac{25}{7}\mathbf{i}_2 + 0\mathbf{i}_3$$

i.e.
$$P \text{ is the point } \left(\frac{10}{7}, \frac{25}{7}, 0\right).$$

Examples VI

1. Find the equation of the lines joining the pairs of points

(i) (2, 3, 5), (1, 0, 7);
(ii) (1, 5, 2), (−3, 2, 4);
(iii) (4, 2, 5), (−1, 1, −2);
(iv) (−5, 1, 2), (4, −3, 1);
(v) (2, 5, −1), (−7, 1, 3).

2. Find the equation of the line which

(i) passes through A (1, −1, 3) and has direction ratios 2 : −1 : 2;
(ii) passes through A (−2, 1, 5) and has direction ratios 2 : 3 : 4;
(iii) passes through A (−1, −2, 1) and has direction ratios 3 : 4 : 6.
(iv) passes through A (2, −3, 4) and has direction ratios 1 : 3 : −2;
(v) passes through A (2, −3, 4) and is normal to the plane BCD where B, C, D are the points (1, 2, 9), (4, 5, 6), (2, −3, 4) respectively;
(vi) passes through A (−1, −2, 4) and is normal to the plane FGH, where F, G, H are the points (−2, 3, −4), (3, 1, 8), (4, −1, 0) respectively.

3. Find the equation of the plane which contains the points

(i) A (1, 1, −3/4),	B (4, −3, −2),	C (2, −3, −3);
(ii) A (2, 1, 2),	B (−2, −1, 3),	C (7, 2, −3);
(iii) A (3, 1, −4),	B (2, −1, 2),	C (−3, 2, 1);
(iv) A (−2, −3, 5),	B (9, 1, 4),	C (7, −3, 2).

4. Find the equation of the plane which passes through the point (1, 3, −4) and is normal to the vector $(2\mathbf{i}_1 + 2\mathbf{i}_2 + \mathbf{i}_3)$.

5. Find the equation of the plane which contains the point (−2, 1, 5) and is normal to the vector $(3\mathbf{i}_1 + 4\mathbf{i}_2 + 6\mathbf{i}_3)$.

6. Find the equation of the plane which contains the point (3, −2, 1) and is normal to the vector $(-2\mathbf{i}_1 + 3\mathbf{i}_2 + 4\mathbf{i}_3)$.

7. Find the equation of the plane which contains the point (1, 1, 8) and is normal to the line

$$\mathbf{r} = 2\mathbf{i}_1 + 3\mathbf{i}_2 + 5\mathbf{i}_3 + t(3\mathbf{i}_1 - 2\mathbf{i}_2 + 2\mathbf{i}_3)$$

8. Find the position vector of the point in which the line joining A (2, 1, 7) and B (−3, 2, 1) cuts the (y, z) plane.

9. Find the position vector of the point of intersection of AB and CD where A, B, C, D are the points whose position vectors are $8\mathbf{i}_1 - 7\mathbf{i}_2 + 2\mathbf{i}_3$, $18\mathbf{i}_1 - 17\mathbf{i}_2 - 3\mathbf{i}_3$, $2\mathbf{i}_1 + 5\mathbf{i}_2 - 3\mathbf{i}_3$, $4\mathbf{i}_1 + 6\mathbf{i}_2 - 8\mathbf{i}_3$.

10. A rectangular box has three adjacent edges OA, OB, OC of lengths 3, 2, 1 units respectively, the lines $\overline{OA}, \overline{OB}, \overline{OC}$ forming a right-handed system of axes. The corners of the box diagonally opposite to O, A, B, C are O', A', B', C' respectively. Taking $\mathbf{i}_1, \mathbf{i}_2, \mathbf{i}_3$ as unit vectors along $\overline{OA}, \overline{OB}, \overline{OC}$ respectively, show that the equation of the plane $A'B'C'$ is

$$\mathbf{r} \,.\, (2\mathbf{i}_1 + 3\mathbf{i}_2 + 6\mathbf{i}_3) = 12$$

Find (i) the perpendicular distance from O to this plane, and (ii) the position vector of the point where OO' cuts the plane $A'B'C'$.

11. Find the perpendicular distance between the point A (4, 3, 3) and the line joining the points B (3, 5, 3) and C (3, 3, 6).

12. O, A, B, C and O', A', B', C' are corresponding corners of opposite faces of a cube of side $2a$; i.e. OO', AA', BB', CC' are parallel edges of the cube. $\overline{OA}$, $\overline{OC}$, $\overline{OO'}$ form a right-handed system of axes, and H, K, are the mid-points of OA, $O'C'$ respectively. Find (1) the perpendicular distance from O' to the plane HKA', (2) the perpendicular distance between the plane HKA' and the plane parallel to it through B.

13. Find the perpendicular distance of the plane through the points (1, 2, −1), (1, 1, 2), (2, 3, 2) from the parallel plane through the point (4, −5, 1).

14. A (**a**), B (**b**), C (**c**) are three points on the surface of a sphere of unit radius, whose centre is at the origin. Show that

(i) $(\mathbf{a} \times \mathbf{b}) . (\mathbf{a} \times \mathbf{c}) = \mathbf{b} . \mathbf{c} - (\mathbf{a} . \mathbf{c})(\mathbf{a} . \mathbf{b})$;
(ii) $(\mathbf{a} \times \mathbf{b}) \times (\mathbf{a} \times \mathbf{c}) = \{\mathbf{a} . (\mathbf{b} \times \mathbf{c})\}\mathbf{a}$.

15. Define the product $\mathbf{u} \times \mathbf{v}$ of two vectors **u**, **v** and show that if **u** is of unit magnitude, then $(\mathbf{u} \times \mathbf{v}) \times \mathbf{u}$ is the component of **v** perpendicular to **u**.

A, B are two fixed points, and **w** is a fixed vector perpendicular to $\overline{AB}$. Describe in geometrical terms the loci defined by

(i) $\overline{AP} . \overline{BP} = 0$, and (ii) $\overline{AP} \times \overline{BP} = \mathbf{w}$.

16. The line joining A (**a**) to B (**b**) cuts the line joining P (**p**) to Q (**q**) at R. Show that the position vector **r** of R satisfies the relations

$$\mathbf{a} \times \mathbf{r} + \mathbf{r} \times \mathbf{b} + \mathbf{b} \times \mathbf{a} = 0$$

and

$$\mathbf{p} \times \mathbf{r} + \mathbf{r} \times \mathbf{q} + \mathbf{q} \times \mathbf{p} = 0$$

If R is the mid-point of both AB, PQ find a relationship between the position vectors of A, B, P, Q and show that $APBQ$ is a parallelogram.

17. A, B, C, D and A', B', C', D' are the corresponding corners of opposite faces of a cube of side $2a$. (Thus AA', BB', CC', DD' are parallel edges and are perpendicular to the faces $ABCD$ and $A'B'C'D'$.) The cube is so lettered that the unit vectors $\mathbf{i}_1, \mathbf{i}_2, \mathbf{i}_3$ along $\overline{AB}$, $\overline{AD}$, $\overline{AA'}$, form a mutually perpendicular right-handed system. X, Y, Z are the mid-points of AD, $D'D$, BC. Express $\overline{AX}$, $\overline{AY}$, $\overline{AZ}$, $\overline{AC}$ in the form $r_1\mathbf{i}_1 + r_2\mathbf{i}_2 + r_3\mathbf{i}_3$.

Find the equation, referred to A as origin, of the plane XYC, and find the perpendicular distance of this plane from Z.

18. (i) Show that the perpendicular distance from the origin, of the plane passing through the points (2, 0, −2), (1, 4, −2), (−1, 2, 2), is $6/\sqrt{93}$ units.

(ii) Find the points A, B at which the straight line through the points (6, −3, −5) and (−3, 12, 10) intersects the planes through the origin, which are perpendicular to $\mathbf{i}_1$ and $\mathbf{i}_3$ respectively, and show that the distance between A and B is $\sqrt{59}$ units.

(iii) Find the point at which the line of (ii) intersects the plane of (i).

19. Prove that the shortest distance, p, between two lines $\mathbf{r} = \mathbf{a} + t\mathbf{b}$ and $\mathbf{r} = \mathbf{a}' + t\mathbf{b}'$, is given by

$$p = \pm \frac{[(\mathbf{a} - \mathbf{a}'), \mathbf{b}, \mathbf{b}']}{|\mathbf{b} \times \mathbf{b}'|}.$$

The edges of a tetrahedron are all of length $2l$. Prove that the shortest distance between a pair of opposite edges is $l\sqrt{2}$.

20. Prove that if $\mathbf{n}$, $\mathbf{a}$ are vectors parallel to the lines ON, OA, then $(\mathbf{n} \times \mathbf{a}) \times \mathbf{n}$ is parallel to the projection of OA on a plane perpendicular to ON.

Two planes are inclined at an angle θ and a straight line makes angles α, β with their normals. Show that if the projections of the line on the two planes are at right-angles then,

$$1 - \cos^2 \alpha - \cos^2 \beta = \pm \cos \alpha \cos \beta \cos \theta.$$

CHAPTER VII

§7.1. Parametric Equations

If $\mathbf{r}$ is the position vector of a point P then $\mathbf{r}$ may be expressed in the form

$$\mathbf{r} = x\mathbf{i}_1 + y\mathbf{i}_2 + z\mathbf{i}_3 \tag{1}$$

where each of the components x, y, z must be known to determine uniquely the position of P. Suppose that each of x, y, z is a one-valued continuous function of a continuous real variable t, so that

$$x = f_1(t),$$
$$y = f_2(t),$$
$$z = f_3(t);$$

then t is the *parameter* of the point P. If T is the set of values which t can take so that $t \epsilon T$, the relation (1) defines a set of points $\{P\}$. Since any particular point P of the set is determined by a particular element t of the set T, the set of points $\{P\}$ is said to have *one degree of freedom*. Geometrically, such a set of points defines *a line*.

Next suppose that each of x, y, z is a one-valued function of the two continuous real variables t, s, the set of values of t being T and the set of values of s being S, so that

$$x = F_1(t, s),$$
$$y = F_2(t, s),$$
$$z = F_3(t, s),$$

where $t \epsilon T$ and $s \epsilon S$. The position vector $\mathbf{r}$ is now determined uniquely only if the values of both the parameters t and s are known; the point P is said to have *two degrees of freedom.* Geometrically, the set of points $\{P\}$ is said to define *a surface.*

We have already in Chapter VI considered the case in which $f_1(t)$, $f_2(t)$, $f_3(t)$ are all expressions of the first degree in t, and the set of points P lies on a straight line. We have also considered the case where $F_1(t, s)$, $F_2(t, s)$, $F_3(t, s)$ are of the first degree in t, s, and the set of points P determines a plane surface. We shall now consider a few simple cases in which the functions $f_k(t)$, $F_k(t, s)$ $(k = 1, 2, 3)$ are not algebraic functions of the first degree.

§7.2. Tangent to a Curve

Suppose the equation of a curve is

$$\mathbf{r} = x\mathbf{i}_1 + y\mathbf{i}_2 + z\mathbf{i}_3, \tag{2}$$

where $x = f_1(t)$, $y = f_2(t)$, $z = f_3(t)$. Let P be a point on the curve with parameter t, and let P' be a point on the curve, near to P, with parameter $t + \delta t$. Then

$$\mathbf{r} + \delta\mathbf{r} = (x + \delta x)\mathbf{i}_1 + (y + \delta y)\mathbf{i}_2 + (z + \delta z)\mathbf{i}_3, \tag{3}$$

therefore $\quad \delta\mathbf{r} = \delta x\, \mathbf{i}_1 + \delta y\, \mathbf{i}_2 + \delta z\, \mathbf{i}_3,$

hence

$$\delta\mathbf{r} \simeq \left(\frac{dx}{dt}\mathbf{i}_1 + \frac{dy}{dt}\mathbf{i}_2 + \frac{dz}{dt}\mathbf{i}_3\right)\delta t. \tag{4}$$

For a given small increment δt of t, there is a unique point P' on the curve; as $\delta t \to 0$, $P' \to P$, and the line PP' tends to the tangent at P. From (4) it can be seen that the direction ratios of PP' are $dx/dt : dy/dt : dz/dt$, i.e. the direction ratios of the tangent at P are $dx/dt : dy/dt : dz/dt$.

Any line through P perpendicular to the tangent at P is normal to the curve, thus there is an infinite number of such normals and these all lie in the plane through P normal to the vector

$$\frac{dx}{dt}\mathbf{i}_1 + \frac{dy}{dt}\mathbf{i}_2 + \frac{dz}{dt}\mathbf{i}_3.$$

§7.3. Tangent Plane to a Surface

Suppose the equation of a surface is given by

$$\mathbf{r} = x\,\mathbf{i}_1 + y\,\mathbf{i}_2 + z\,\mathbf{i}_3 \tag{5}$$

where $x = F_1(t, s)$, $y = F_2(t, s)$, $z = F_3(t, s)$. Let P be a point on the surface whose position vector is $\mathbf{r}$ and parameters t, s; let P' be a neighbouring point on the surface whose position vector is $\mathbf{r} + \delta\mathbf{r}$ and whose parameters are $t + \delta t$, $s + \delta s$. Then with the usual notation

$$\delta\mathbf{r} = \delta x\,\mathbf{i}_1 + \delta y\,\mathbf{i}_2 + \delta z\,\mathbf{i}_3 \tag{6}$$

hence

$$\delta\mathbf{r} \simeq \left(\frac{\partial x}{\partial t}\,\delta t + \frac{\partial x}{\partial s}\,\delta s\right)\mathbf{i}_1 + \left(\frac{\partial y}{\partial t}\,\delta t + \frac{\partial y}{\partial s}\,\delta s\right)\mathbf{i}_2 + \left(\frac{\partial z}{\partial t}\,\delta t + \frac{\partial z}{\partial s}\,\delta s\right)\mathbf{i}_3 \tag{7}$$

i.e.

$$\delta\mathbf{r} \simeq \left(\frac{\partial x}{\partial t}\,\mathbf{i}_1 + \frac{\partial y}{\partial t}\,\mathbf{i}_2 + \frac{\partial z}{\partial t}\,\mathbf{i}_3\right)\delta t + \left(\frac{\partial x}{\partial s}\,\mathbf{i}_1 + \frac{\partial y}{\partial s}\,\mathbf{i}_2 + \frac{\partial z}{\partial s}\,\mathbf{i}_3\right)\delta s. \tag{8}$$

Since δt, δs, are independent small increments, the relation (7) defines a set of lines PP' all of which become tangents to the surface (5) as $\delta t \to 0$, $\delta s \to 0$. Let λ, μ, ν, be chosen so that

$$\left.\begin{aligned} \lambda\,\frac{\partial x}{\partial t} + \mu\,\frac{\partial y}{\partial t} + \nu\,\frac{\partial z}{\partial t} &= 0 \\ \text{and} \qquad \lambda\,\frac{\partial x}{\partial s} + \mu\,\frac{\partial y}{\partial s} + \nu\,\frac{\partial z}{\partial s} &= 0 \end{aligned}\right\} \tag{9}$$

Then the vector $\mathbf{n} = \lambda\mathbf{i}_1 + \mu\mathbf{i}_2 + \nu\mathbf{i}_3$ is perpendicular to PP' for all points P' determined by (7) since

$$\mathbf{n}\,.\,\delta\mathbf{r} = \left(\lambda\frac{\partial x}{\partial t} + \mu\frac{\partial y}{\partial t} + \nu\frac{\partial z}{\partial t}\right)\delta t + \left(\lambda\frac{\partial x}{\partial s} + \mu\frac{\partial y}{\partial s} + \nu\frac{\partial z}{\partial s}\right)\delta s$$

$$= 0,$$

i.e. all the tangents through P to the surface (5) are normal to the vector $\mathbf{n}$. Hence all the tangents at P to the surface (5) lie in the plane through P whose normal has direction ratios $\lambda : \mu : \nu$. This plane is known as the tangent plane to the surface (5). The ratios $\lambda : \mu : \nu$ are equal to the ratios of the second order determinants of the matrix

$$\begin{pmatrix} \dfrac{\partial x}{\partial t} & \dfrac{\partial y}{\partial t} & \dfrac{\partial z}{\partial t} \\ \dfrac{\partial x}{\partial s} & \dfrac{\partial y}{\partial s} & \dfrac{\partial z}{\partial s} \end{pmatrix}. \tag{10}$$

Hence the tangent plane to the surface (5) at the point whose parameters are t, s is the plane through that point whose normal has direction ratios given by the ratios of the second order determinants of the matrix (10).

Example 1. *Find the equation of the tangent at the point* $t = \pi/3$ *to the curve* $\mathbf{r} = (\log \cos t)\mathbf{i}_1 + (\log \sin t)\mathbf{i}_2 + (t\sqrt{3})\mathbf{i}_3$.

The given curve is

$$\mathbf{r} = (\log \cos t)\mathbf{i}_1 + (\log \sin t)\mathbf{i}_2 + (t\sqrt{3})\mathbf{i}_3.$$

Then the direction ratios of the tangent at the point P whose parameter is t are

$$\frac{1}{\cos t_1}(-\sin t_1) : \frac{1}{\sin t_1}(\cos t_1) : \sqrt{3},$$

i.e.

$$(-\tan t_1) : \frac{1}{\tan t_1} : \sqrt{3}.$$

Hence the direction ratios of the tangent at the point A whose parameter is $\pi/3$ are

$$(-\sqrt{3}) : \frac{1}{\sqrt{3}} : \sqrt{3},$$

i.e.

$$(-3) : 1 : 3,$$

therefore the equation of the tangent at A is

$$\mathbf{r} = (\log 1/2)\mathbf{i}_1 + (\log \sqrt{3}/2)\mathbf{i}_2 + (\pi\sqrt{3}/3)\mathbf{i}_3 + (-3\mathbf{i}_1 + \mathbf{i}_2 + 3\mathbf{i}_3)s$$

where s is a parameter,

i.e. $$\mathbf{r} = \left(-3s + \log\frac{1}{2}\right)\mathbf{i}_1 + \left(s + \log\frac{\sqrt{3}}{2}\right)\mathbf{i}_2 + \left(3s + \frac{\pi\sqrt{3}}{3}\right)\mathbf{i}_3.$$

Example 2. *Find the equation of the plane which touches the surface* $\mathbf{r} = (t + 2s)\mathbf{i}_1 + (t - 2s)\mathbf{i}_2 + 4ts\mathbf{i}_3$ *at the point where it is cut by the line* $\mathbf{r} = (5 - \theta)\mathbf{i}_1 + (1 - \theta)\mathbf{i}_2 + (12 - \theta)\mathbf{i}_3$ *where* t, s, θ *are parameters.*

Suppose the line

$$\mathbf{r} = (5 - \theta)\mathbf{i}_1 + (1 - \theta)\mathbf{i}_2 + (12 - \theta)\mathbf{i}_3 \tag{i}$$

cuts the surface

$$\mathbf{r} = (t + 2s)\mathbf{i}_1 + (t - 2s)\mathbf{i}_2 + 4ts\mathbf{i}_3 \tag{ii}$$

at the point P_1 whose parameters are t, s, θ_1.

Then

$$5 - \theta_1 = t_1 + 2s_1 \tag{iii}$$

$$1 - \theta_1 = t_1 - 2s_1 \tag{iv}$$

$$12 - \theta_1 = 4t_1s_1 \tag{v}$$

From (iii) and (iv), $s_1 = 1,\ 3 - t_1 = \theta_1$

and from (v) $$12 - \theta_1 = 4(3 - \theta_1)(1)$$

$$\theta_1 = 0$$

and $$t_1 = 3.$$

Hence (i) and (ii) intersect at the point P_1 (5, 1, 12).

From (ii) we see that the direction ratios of the normal to the surface at the point P are $\lambda : \mu : \nu$ where

$$(1)\lambda + (1)\mu + (4s)\nu = 0,$$

$$(2)\lambda + (-2)\mu + (4t)\nu = 0.$$

Hence the direction ratios of the normal to the surface at P_1 are given by

$$\begin{cases} \lambda + \mu + 4\nu = 0 \\ 2\lambda - 2\mu + 12\nu = 0 \end{cases}$$

i.e.

$$\begin{cases} \lambda + \mu + 4\nu = 0 \\ \lambda - \mu + 6\nu = 0 \end{cases}$$

therefore $$\frac{\lambda}{10} = \frac{\mu}{-2} = \frac{\nu}{-2}$$

i.e. $$\frac{\lambda}{5} = \frac{\mu}{-1} = \frac{\nu}{-1}$$

therefore the equation of the tangent plane at P_1 is

$$\mathbf{r}\,(5\mathbf{i}_1 - \mathbf{i}_2 - \mathbf{i}_3) = (5\mathbf{i}_1 + \mathbf{i}_2 + 12\mathbf{i}_3)\ .\ (5\mathbf{i}_1 - \mathbf{i}_2 - \mathbf{i}_3)$$

i.e. $$\mathbf{r}\ .\ (5\mathbf{i}_1 - \mathbf{i}_2 - \mathbf{i}_3) = 12.$$

§7.4. Some Special Cases

(a) Consider the curve whose equation is

$$\mathbf{r} = at^2\mathbf{i}_1 + 2at\mathbf{i}_2. \tag{I}$$

Since there is no component in the direction of $\mathbf{i}_3$, the vector $\mathbf{r}$ must lie in the plane of $\mathbf{i}_1$ and $\mathbf{i}_2$. Hence the curve (I) must be a plane curve lying in the (x, y) plane. Also, since it has parametric coordinates $x = at^2$, $y = 2at$, we see that the curve is a parabola with OX as axis, O as vertex, and OY as the tangent at the vertex.

(b) Similarly the curve

$$\mathbf{r} = (a \cos \theta)\mathbf{i}_1 + (b \sin \theta)\mathbf{i}_2 \tag{II}$$

can be seen to be an ellipse in the (x, y) plane having its major axis of length $2a$ along the axis OX, its minor axis of length $2b$ along OY, and its centre at O.

(c) The surface

$$\mathbf{r} = (a \cos \theta)\mathbf{i}_1 + (b \sin \theta)\mathbf{i}_2 + c\phi\mathbf{i}_3 \tag{III}$$

can be seen to be a cylinder with its axis along OZ, any cross-section parallel to the (x, y) plane being an ellipse with its centre on OZ, and axes of lengths $2a$ and $2b$ parallel to OX and OY respectively.

(d) The equation

$$\mathbf{r} = (a \cos \theta)\mathbf{i}_1 + (b \sin \theta)\mathbf{i}_2 + c\theta\mathbf{i}_3 \tag{IV}$$

represents a curve since it has only one degree of freedom. Moreover, the set of points which define this curve is the sub-set of the points which define the surface (III) for which $\theta = \phi$. Hence the curve (IV) is a curve described on the surface of the cylinder (III).

(e) If in equation (III) above we put $a = b$, the surface becomes the circular cylinder whose equation is

$$\mathbf{r} = (a \cos \theta)\mathbf{i}_1 + (a \sin \theta)\mathbf{i}_2 + c\phi\mathbf{i}_3. \tag{V}$$

(f) If, in equation (IV) above, $a = b$, the equation of the curve becomes

$$\mathbf{r} = (a \cos \theta)\mathbf{i}_1 + (a \sin \theta)\mathbf{i}_2 + c\theta\mathbf{i}_3 \qquad \text{(VI)}$$

which represents a curve described on the surface of a circular cylinder of radius a.

From equation (VI) we see that the direction ratios of the tangent at the point P whose parameter is θ are

$$(- a \sin \theta) : (a \cos \theta) : c.$$

therefore the direction cosines of the tangent at P are

$$-\frac{a \sin \theta}{\sqrt{(a^2 + c^2)}}, \quad \frac{a \cos \theta}{\sqrt{(a^2 + c^2)}}, \quad \frac{c}{\sqrt{(a^2 + c^2)}}.$$

From this result it can be seen that the tangent at P is inclined to the z-axis at a constant angle $\cos^{-1} c/\sqrt{(a^2 + c^2)}$. Any curve with the property that the tangent makes a constant angle with a fixed direction is a *helix*. Thus the curve represented by equation (VI) is a helix; since this curve is described on the surface of a circular cylinder it is known as a *circular helix*.

Examples VII

1. Show that the tangent at any point of the helix

$$\mathbf{r} = (a \cos \theta)\mathbf{i}_1 + (a \sin \theta)\mathbf{i}_2 + a\theta\mathbf{i}_3$$

makes an angle $\pi/4$ with OZ. Find the equation of the circular helix which makes a constant angle $\pi/3$ with OX.

2. Show that the parabolas

$$\mathbf{r} = 4t^2\mathbf{i}_1 + 2t\mathbf{i}_2$$

and

$$\mathbf{r} = (4 + s^2)\mathbf{i}_1 + 2\mathbf{i}_2 + 2s\mathbf{i}_3$$

intersect. Find their point of intersection, and the angle between their tangents at this point.

3. Find the points of intersection of the circle

$$\mathbf{r} = (4 \cos \theta)\mathbf{i}_1 + (4 \sin \theta)\mathbf{i}_2$$

and the parabola

$$\mathbf{r} = 2t\mathbf{i}_1 + (t^2 - 4)\mathbf{i}_3.$$

4. Show that the curve

$$\mathbf{r} = (a \cosh t \cos t)\mathbf{i}_1 + (a \cosh t \sin t)\mathbf{i}_2 + at\mathbf{i}_3$$

touches the helix

$$\mathbf{r} = (a \cos \theta)\mathbf{i}_1 + (a \sin \theta)\mathbf{i}_2 + a\theta\mathbf{i}_3.$$

Find the position vector of the point of contact, and the equation of their common tangent.

5. Show that the curve

$$\mathbf{r} = (a_1t^2 + b_1t + c_1)\mathbf{i}_1 + (a_2t^2 + b_2t + c_2)\mathbf{i}_2 + (a_3t^2 + b_3t + c_3)\mathbf{i}_3$$

is a plane curve and find the equation of the plane in which it lies.

6. Show that a circular helix cuts off equal intercepts on any generator of the cylinder on which it is described.

7. Show that the helices

$$\mathbf{r} = (a \cos \theta)\mathbf{i}_1 + (a \sin \theta)\mathbf{i}_2 + a\theta\mathbf{i}_3$$

and

$$\mathbf{r} = (a \cos \phi)\mathbf{i}_1 + (a \sin \phi)\mathbf{i}_2 + 2a\phi\mathbf{i}_3$$

have an infinite number of points of intersection. Show also that all these points of intersection lie on the same line parallel to OZ. If P_1, P_2 ... are successive points of intersection, show that the distances P_1P_2, P_2P_3 ... are equal; find this common distance.

8. If A is the point $t = \pi/4$ on the curve

$$\mathbf{r} = (\log \cos t)\mathbf{i}_1 + (\log \sin t)\mathbf{i}_2 + (t\sqrt{2})\mathbf{i}_3,$$

and if OB is the perpendicular from the origin to the tangent at A, find the coordinates of B.

9. Show that the curve

$$\mathbf{r} = at^3\mathbf{i}_1 + a\left(\sqrt{\frac{3}{2}}\right)t^2\mathbf{i}_2 + at\mathbf{i}_3$$

cuts the surface

$$\mathbf{r} = (a \cos \theta \cos \phi)\mathbf{i}_1 + (a \cos \theta \sin \phi)\mathbf{i}_2 + (a \sin \theta)\mathbf{i}_3$$

at points whose parameter t satisfies the equation

$$2t^6 + 3t^4 + 2t^2 - 2 = 0.$$

Show that one solution of this equation is $t = 1/\sqrt{2}$, and find the corresponding values of θ, ϕ. If A is the point for which $t = 1/\sqrt{2}$, find the angle between the curve and the surface at A.

(N.B. *The angle between a curve and a surface at a point of intersection is the angle between the tangent to the curve and the tangent plane to the surface.*)

ANSWERS

Examples Ia Page 9

6. Radius of locus of $Q = 4$ (Radius of locus of P).
7. B moves on the arc XY remote from A.

Examples Ib Page 20

1. $\overline{PQ} = -5\mathbf{i}_1 - 8\mathbf{i}_2 + 5\mathbf{i}_3$; dir. rat. $5 : 8 : -5$; magnitude $\sqrt{114}$.
2. $\overline{AB} = 8\mathbf{i}_1 + 3\mathbf{i}_2 + 5\mathbf{i}_3$; dir. rat. $8 : 3 : 5$; magnitude $7\sqrt{2}$.
3. $\overline{AC} = \mathbf{a} + 3\mathbf{b}$; $\overline{DB} = -\mathbf{a} + 3\mathbf{b}$; $\overline{BC} = 2\mathbf{a} + 2\mathbf{b}$; $\overline{CA} = -\mathbf{a} - 3\mathbf{b}$.
4. $D \ldots \mathbf{a} - 2\mathbf{b} + 2\mathbf{c}$; $E \ldots 2\mathbf{a} - 3\mathbf{b} + 2\mathbf{c}$; $F \ldots 2\mathbf{a} - 2\mathbf{b} + \mathbf{c}$.
14. Dir. rat. $3 : 10 : 3$; $OC = \frac{1}{5}\sqrt{118}$.
18. Resultant $= 8\mathbf{i}_1 + 13\mathbf{i}_2 + 14\mathbf{i}_3$ lb wt.; dir. rat. $8 : 13 : 14$; magnitude 20·7 lb wt.
19. Resultant = 36·38 dyn; dir. rat. $5 : 9 : 15$.
20. 7·42 lb wt.; dir. rat. $5 : 6 : 7$.
21. 7·18 lb wt.; dir. rat. $3(\sqrt{2} + \sqrt{3}) : 3(\sqrt{2} + 2\sqrt{3}) : (3\sqrt{2} - \sqrt{3})$.

Examples IIa Page 38

1. $+3$. **2.** $+2$. **3.** -3. **4.** -17.
5. $19\mathbf{i}_1 - 8\mathbf{i}_2 - 10\mathbf{i}_3$. **6.** $19\mathbf{i}_1 + 11\mathbf{i}_2 - \mathbf{i}_3$.
7. $-28\mathbf{i}_1 + 20\mathbf{i}_2 + 4\mathbf{i}_3$. **8.** $-5\mathbf{i}_1 + 23\mathbf{i}_2 + 3\mathbf{i}_3$.
9. -20 ft lb. **10.** $(-27\mathbf{i}_1 + 24\mathbf{i}_2 - 3\mathbf{i}_3)$ dyn cm units.
11. (i) -23; (ii) $\mathbf{i}_1 - 6\mathbf{i}_2 - 32\mathbf{i}_3$.
12. (i) -14; (ii) $-45\mathbf{i}_1 + 12\mathbf{i}_2 - \mathbf{i}_3$.
13. $26/\sqrt{11}$ ft lb.
14. $(40\mathbf{i}_1 - 22\mathbf{i}_2 - 2\mathbf{i}_3)$ lb ft units.
15. $\hat{A} = \cos^{-1}23/(\sqrt{41})(\sqrt{19})$; $\hat{B} = \cos^{-1}18/(\sqrt{14})(\sqrt{41})$;
$\hat{C} = \cos^{-1}(-4/(\sqrt{14})(\sqrt{19})$.

Examples IIb Page 43

1. (i) 91; (ii) 132; (iii) -160; (iv) -25; (v) 103.
2. (i) $-94\mathbf{i}_1 - 72\mathbf{i}_2 + 62\mathbf{i}_3$; (ii) $-86\mathbf{i}_1 - 27\mathbf{i}_2 - 34\mathbf{i}_3$;
(iii) $56\mathbf{i}_1 + 32\mathbf{i}_2 + 96\mathbf{i}_3$; (iv) $27\mathbf{i}_1 + 206\mathbf{i}_2 + 134\mathbf{i}_3$;
(v) $45\mathbf{i}_1 - 113\mathbf{i}_2 - 57\mathbf{i}_3$.
3. (i) 113; $-6\mathbf{i}_1 - 15\mathbf{i}_2 + 12\mathbf{i}_3$ (ii) 44; $-42\mathbf{i}_1 - 72\mathbf{i}_2 + 54\mathbf{i}_3$
(iii) -51; $-59\mathbf{i}_1 + 23\mathbf{i}_2 - 26\mathbf{i}_3$ (iv) 86; $-12\mathbf{i}_1 - \mathbf{i}_2 + 8\mathbf{i}_3$
(v) -92; $-731\mathbf{i}_1 + 558\mathbf{i}_2 - 331\mathbf{i}_3$.

Examples III Page 52

1. (i) $(2\cos t)\mathbf{i}_1 - (3\sin t)\mathbf{i}_2 + (2\cos 2t)\mathbf{i}_3$;
(ii) $(12t^3)\mathbf{i}_1 + (15t^2)\mathbf{i}_2 + 4\mathbf{i}_3$;

(iii) $(2e^{2t})\mathbf{i}_1 + (3e^{3t})\mathbf{i}_2 + (4e^{4t})\mathbf{i}_3$;
(iv) $(6t)\mathbf{i}_1 + (3t^2 - 8t + 3)\mathbf{i}_2 + (4\cos 4t)\mathbf{i}_3$;
(v) $\{t/\sqrt{(t^2+1)}\}\mathbf{i}_1 + (1/t)\mathbf{i}_2 + (12t^3)\mathbf{i}_3$.

2. (i) $(2 + 3t^2)\mathbf{i}_1 + (6t + 1)\mathbf{i}_2 + (12t^2)\mathbf{i}_3$;
(ii) $4 - 18t + 60t^2$;
(iii) $(9t^2 - 4)\mathbf{i}_1 + (18t^2)\mathbf{i}_2 + (2 + 9t^2)\mathbf{i}_3$;
(iv) $-(15t^4 + 20t^3)\mathbf{i}_1 + (30t^5 + 8t^3)\mathbf{i}_2 + (-15t^4 + 4t)\mathbf{i}_3$;
(v) $(-3 + 3t^2 + 8t^3 - 30t^5)$;
(vi) $(16t - 64t^3 - 90t^4 - 90t^5)$;
(vii) $(-12t - 32t^3 + 45t^4 - 120t^5)\mathbf{i}_1 + (-8t - 80t^3 - 30t^5)\mathbf{i}_2$
$+ (-27t^2 - 44t^3 + 60t^5)\mathbf{i}_3$;
(viii) $8t + 36t^3 + 150t^5$;
(ix) $9t^2 + 8t^3 - 30t^5$;
(x) $2t + 12t^5$.

In the following solutions **k** is a constant vector and c a constant scalar.

4. (i) $\frac{t^3}{3}\mathbf{i}_1 + \frac{t^2}{2}\mathbf{i}_2 + t\mathbf{i}_3 + \mathbf{k}$;
(ii) $(-\cos t)\mathbf{i}_1 + (\sin t)\mathbf{i}_2 + \frac{t^2}{2}\mathbf{i}_3 + \mathbf{k}$;
(iii) $e^t\mathbf{a} + \mathbf{k}$; (iv) $\frac{t^2}{2} + \frac{t^4}{2} + c$;

6. (i) $(te^t)\mathbf{a} \cdot \mathbf{b} + c$;
(ii) $(-\frac{1}{2}\sin 2t)\mathbf{a} \times \mathbf{b} + \mathbf{k}$.

8. $(9t^2)\mathbf{i}_1 + (1/t)\mathbf{i}_2 - (1/t^2)\mathbf{i}_3$.

9. (i) $3(t^2 - 1)\mathbf{i}_1 + 4t^3\mathbf{i}_2 + 3(t^2 + 1)\mathbf{i}_3$; (ii) $-2t$.

Examples IVa Page 59

1. (i) $(\cos t)\mathbf{i}_1 - (\sin t)\mathbf{i}_2 + \mathbf{i}_3$; $-(\sin t)\mathbf{i}_1 - (\cos t)\mathbf{i}_2$;
(ii) $3t^2\mathbf{i}_1 + 2t\mathbf{i}_2 + 3\mathbf{i}_3$; $6t\mathbf{i}_1 + 2\mathbf{i}_2$;
(iii) $\left(\frac{1}{t}\right)\mathbf{i}_1 - 2e^{2t}\mathbf{i}_2 + 2t\mathbf{i}_3$; $-\left(\frac{1}{t^2}\right)\mathbf{i}_1 - 4e^{2t}\mathbf{i}_2 + 2\mathbf{i}_3$;
(iv) $2\mathbf{a}e^{2t}$; $4\mathbf{a}e^{2t}$;
(v) $2\mathbf{a}\cos 2t - 2\mathbf{b}\sin 2t$; $-4\mathbf{a}\sin 2t - 4\mathbf{b}\cos 2t$.

2. Velocity $= a\omega e^{\theta}\mathbf{l}_1 + a\omega e^{\theta}\mathbf{l}_2$;
Acceleration $= 2a\omega^2 e^{\theta}\mathbf{l}_2$.

3. Velocity $= \{\frac{1}{3}(r^2\omega \sin\theta)\mathbf{l}_1 + r\omega\mathbf{l}_2$;
Acceleration $= \frac{1}{3}r^2\omega^2(2 - \cos\theta) - r\omega^2\}\mathbf{l}_1 + \frac{2}{3}r^2\omega^2(\sin\theta)\mathbf{l}_2$

4. (i) $\dot{\mathbf{r}} = a(1 - \cos t)\mathbf{i}_1 + (a\sin t)\mathbf{i}_2 + b\mathbf{i}_3$;
$\ddot{\mathbf{r}} = (a\sin t)\mathbf{i}_1 + (a\cos t)\mathbf{i}_2$;
(ii) $\dot{\mathbf{r}} = (-a\sin t)\mathbf{i}_1 + (a\cos t)\mathbf{i}_2 - (2a\sin 2t)\mathbf{i}_3$;
$\ddot{\mathbf{r}} = (-a\cos t)\mathbf{i}_1 - (a\sin t)\mathbf{i}_2 - (4a\cos 2t)\mathbf{i}_3$.

5. $\dot{\mathbf{r}} = ak\mathbf{l}_1 + ak^2t\mathbf{l}_2;$
$\ddot{\mathbf{r}} = -ak^3t\mathbf{l}_1 + 2ak^2\mathbf{l}_2.$

6. (i) $\dot{\mathbf{r}} = -\dfrac{a(t^2-1)}{(t^2+1)^2}\mathbf{l}_1 + \dfrac{a(t^2-1)}{t(t^2+1)}\mathbf{l}_2;$

$$\ddot{\mathbf{r}} = a\left\{\frac{2t(t^2-3)}{(t^2+1)^3} - \frac{(t^2-1)^2}{t^3(t^2+1)}\right\}\mathbf{l}_1 + \frac{2a}{t^2}\left\{\frac{1}{(t^2+1)} - \frac{(t^2-1)^2}{(t^2+1)^2}\right\}\mathbf{l}_2.$$

(ii) $\dot{\mathbf{r}} = -\tfrac{1}{2}\left(\dfrac{2}{t}\right)^{3/2}\mathbf{l}_1 + \left(\dfrac{2}{t}\right)^{1/2}\mathbf{l}_2;$

$$\ddot{\mathbf{r}} = \left\{\frac{3}{8}\left(\frac{2}{t}\right)^{5/2} - \frac{1}{2}\left(\frac{2}{t}\right)^{\frac{1}{2}}\right\}\mathbf{l}_1 - \tfrac{1}{2}\left(\frac{2}{t}\right)^{3/2}\mathbf{l}_2.$$

(iii) $\dot{\mathbf{r}} = e^{-t}\mathbf{l}_1 + (e^t - 1)\mathbf{l}_2;$
$\ddot{\mathbf{r}} = -(e^{2t} - e^t + e^{-t})\mathbf{l}_1 + (e^t + 1)\mathbf{l}_2.$

(iv) $\dot{\mathbf{r}} = (2\sin 2t)\mathbf{l}_1 + 2(4 - \cos 2t)\mathbf{l}_2;$
$\ddot{\mathbf{r}} = -8(2 - \cos 2t)\mathbf{l}_1 + (8\sin 2t)\mathbf{l}_2.$

7. (i) $\dot{\mathbf{r}} = -\dfrac{\sin\theta}{6}\mathbf{l}_1 + \dfrac{2-\cos\theta}{6}\mathbf{l}_2;$

$\ddot{\mathbf{r}} = -\dfrac{(2-\cos\theta)^2}{108}\mathbf{l}_1.$

(ii) $\dot{\mathbf{r}} = \left(\dfrac{2}{5}\sin\theta\right)\mathbf{l}_1 + \left(\dfrac{3 + 2\cos\theta}{5}\right)\mathbf{l}_2;$

$\mathbf{r} = -\dfrac{3}{125}\left(3 + 2\cos\theta\right)^2\mathbf{l}_1.$

8. (i) $\dot{\mathbf{r}} = 6t\mathbf{b};$ $\quad\ddot{\mathbf{r}} = 6\mathbf{b};$
(ii) $\dot{\mathbf{r}} = 4\mathbf{b};$ $\quad\ddot{\mathbf{r}} = 0;$
(iii) $\dot{\mathbf{r}} = (\cos t)\mathbf{b};$ $\quad\ddot{\mathbf{r}} = (-\sin t)\mathbf{b};$
(iv) $\dot{\mathbf{r}} = \left(1 + \dfrac{1}{t}\right)\mathbf{b};$ $\ddot{\mathbf{r}} = -\dfrac{1}{t^2}\mathbf{b}.$

Examples IVb Page 64

1. Velocity $= (5t + 16)\mathbf{i}_1 + (\tfrac{5}{2}t + 16)\mathbf{i}_2 + (3t + 28)\mathbf{i}_3.$
Position vector w.r.t. point of projection,
$$(\tfrac{5}{2}t^2 + 16t)\mathbf{i}_1 + (\tfrac{5}{4}t^2 + 16t)\mathbf{i}_2 + (\tfrac{3}{2}t^2 + 28t)\mathbf{i}_3.$$
2. Velocity $= (10 - 2\cos t)\mathbf{i}_1 + (12 - 64t)\mathbf{i}_2 + 24\mathbf{i}_3.$
Position vector w.r.t. point of projection,
$$(10t - 2\sin t)\mathbf{i}_1 + (12t - 32t^2)\mathbf{i}_2 + 24t\mathbf{i}_3.$$
3. Velocity $= (\tfrac{1}{2}t^2 + 16)\mathbf{i}_1 + (49 - \cos t)\mathbf{i}_2 + (72 - 32t)\mathbf{i}_3.$
Position vector w.r.t. point of projection,
$$(\tfrac{1}{6}t^3 + 16t)\mathbf{i}_1 + (49t - \sin t)\mathbf{i}_2 + (72t - 16t^2)\mathbf{i}_3.$$
4. Velocity $= 25\mathbf{i}_2 + (60 - 32t)\mathbf{i}_3.$
Position vector $= 25t\mathbf{i}_2 + (60t - 16t^2)\mathbf{i}_3.$
$3\tfrac{3}{4}$ sec; $93\tfrac{3}{4}$ ft.

5. (i) 2 sec.; (ii) $64\sqrt{3}$ ft.; (iii) 16 ft.

6. 1,000,000 ft.

10. $\mathbf{r} = \left(80 - \frac{2g}{3}\right)\mathbf{n} + \frac{80}{\sqrt{3}}\mathbf{a}$, where **n**, **a** are unit vectors in the plane of motion, **n** being vertical and **a** horizontal.

Examples V Page 77

1. Ellipse; eccentricity $\frac{1}{6}\sqrt{21}$.

2. Eccentricity $= \frac{1}{\sqrt{2}}$; Periodic time $= 2\pi a\sqrt{(a/\mu)}$.

3. Orbit; $2/p^2 = 2 - \log r$.

11. Velocity $= \left\{\left(n^2a^2 - u^2\right) \sin^2 nt + u^2\right\}^{\frac{1}{2}}$; periodic time $= 2\pi/n$.

Examples VI Page 91

1. (i) $\mathbf{r} = 2\mathbf{i}_1 + 3\mathbf{i}_2 + 5\mathbf{i}_3 - t(\mathbf{i}_1 + 3\mathbf{i}_2 - 2\mathbf{i}_3)$;
(ii) $\mathbf{r} = \mathbf{i}_1 + 5\mathbf{i}_2 + 2\mathbf{i}_3 - t(4\mathbf{i}_1 + 3\mathbf{i}_2 - 2\mathbf{i}_3)$;
(iii) $\mathbf{r} = 4\mathbf{i}_1 + 2\mathbf{i}_2 + 5\mathbf{i}_3 - t(5\mathbf{i}_1 + \mathbf{i}_2 + 7\mathbf{i}_3)$;
(iv) $\mathbf{r} = -5\mathbf{i}_1 + \mathbf{i}_2 + 2\mathbf{i}_3 + t(9\mathbf{i}_1 - 4\mathbf{i}_2 - \mathbf{i}_3)$;
(v) $\mathbf{r} = 2\mathbf{i}_1 + 5\mathbf{i}_2 - \mathbf{i}_3 - t(9\mathbf{i}_1 + 4\mathbf{i}_2 - 4\mathbf{i}_3)$.

2. (i) $\mathbf{r} = \mathbf{i}_1 - \mathbf{i}_2 + 3\mathbf{i}_3 + t(2\mathbf{i}_1 - \mathbf{i}_2 + 2\mathbf{i}_3)$;
(ii) $\mathbf{r} = -2\mathbf{i}_1 + \mathbf{i}_2 + 5\mathbf{i}_3 + t(2\mathbf{i}_1 + 3\mathbf{i}_2 + 4\mathbf{i}_3)$;
(iii) $\mathbf{r} = -\mathbf{i}_1 - 2\mathbf{i}_2 + \mathbf{i}_3 + t(3\mathbf{i}_1 + 4\mathbf{i}_2 + 6\mathbf{i}_3)$;
(iv) $\mathbf{r} = 2\mathbf{i}_1 - 3\mathbf{i}_2 + 4\mathbf{i}_3 + t(\mathbf{i}_1 + 3\mathbf{i}_2 - 2\mathbf{i}_3)$;
(v) $\mathbf{r} = 2\mathbf{i}_1 - 3\mathbf{i}_2 + 4\mathbf{i}_3 + t(5\mathbf{i}_1 - 2\mathbf{i}_2 + 3\mathbf{i}_3)$;
(vi) $\mathbf{r} = -\mathbf{i}_1 - 2\mathbf{i}_2 + 4\mathbf{i}_3 + t(10\mathbf{i}_1 + 13\mathbf{i}_2 - 2\mathbf{i}_3)$.

3. (i) $\mathbf{r} = \mathbf{i}_1 + \mathbf{i}_2 - \frac{3}{4}\mathbf{i}_3 + s(3\mathbf{i}_1 - 4\mathbf{i}_2 - \frac{5}{4}\mathbf{i}_3) + t(\mathbf{i}_1 - 4\mathbf{i}_2 - \frac{9}{4}\mathbf{i}_3)$;
(ii) $\mathbf{r} = 2\mathbf{i}_1 + \mathbf{i}_2 + 2\mathbf{i}_3 + s(-4\mathbf{i}_1 - 2\mathbf{i}_2 + \mathbf{i}_3) + t(5\mathbf{i}_1 + \mathbf{i}_2 - 5\mathbf{i}_3)$;
(iii) $\mathbf{r} = 3\mathbf{i}_1 + \mathbf{i}_2 - 4\mathbf{i}_3 + s(-\mathbf{i}_1 - 2\mathbf{i}_2 + 6\mathbf{i}_3) + t(-6\mathbf{i}_1 + \mathbf{i}_2 + 5\mathbf{i}_3)$;
(iv) $\mathbf{r} = -2\mathbf{i}_1 - 3\mathbf{i}_2 + 5\mathbf{i}_3 + s(11\mathbf{i}_1 + 4\mathbf{i}_2 - \mathbf{i}_3) + t(9\mathbf{i}_1 - 3\mathbf{i}_3)$.

4. $\mathbf{r} \cdot (2\mathbf{i}_1 + 2\mathbf{i}_2 + \mathbf{i}_3) = 4$.
5. $\mathbf{r} \cdot (3\mathbf{i}_1 + 4\mathbf{i}_2 + 6\mathbf{i}_3) = 28$.
6. $\mathbf{r} \cdot (-2\mathbf{i}_1 + 3\mathbf{i}_2 + 4\mathbf{i}_3) = -8$.
7. $\mathbf{r} \cdot (3\mathbf{i}_1 - 2\mathbf{i}_2 + 2\mathbf{i}_3) = 17$.
8. $(7/5)\mathbf{i}_2 + (23/5)\mathbf{i}_3$.
9. $-2\mathbf{i}_1 + 3\mathbf{i}_2 + 7\mathbf{i}_3$.
10. (i) 12/7 units; (ii) $2\mathbf{i}_1 + \frac{4}{3}\mathbf{i}_2 + \frac{2}{3}\mathbf{i}_3$.
11. $7/\sqrt{13}$.
12. (i) $4a/\sqrt{21}$; (ii) $10a/\sqrt{21}$.
13. $37/\sqrt{46}$.
17. $\overline{AX} = a\mathbf{i}_2$; $\overline{AY} = 2a\mathbf{i}_2 + a\mathbf{i}_3$; $\overline{AZ} = 2a\mathbf{i}_1 + a\mathbf{i}_2$; $\overline{AC} = 2a\mathbf{i}_1 + 2a\mathbf{i}_2$.
Plane $XYC : \mathbf{r} = a\mathbf{i}_2 + s(a\mathbf{i}_2 + a\mathbf{i}_3) + t(2a\mathbf{i}_1 + a\mathbf{i}_2)$.
Distance: $2a/3$.

18. (iii) $(9, -8, -10)$.

Examples VII Page 101

1. $r = \dfrac{a}{\sqrt{3}}\,\theta\mathbf{i}_1 + (a\cos\theta)\mathbf{i}_2 + (a\sin\theta)\mathbf{i}_3.$

2. (4, 2, 0); $\pi/2$.

3. $(\pm\, 4,\ 0,\ 0)$.

4. $\mathbf{r} = a\mathbf{i}_1$;

   $\mathbf{r} = a\mathbf{i}_1 + s\mathbf{i}_2 + s\mathbf{i}_3.$

5. $\mathbf{r}\,.\,\{(a_2b_3 - a_3b_2)\mathbf{i}_1 + (a_3b_1 - a_1b_3)\mathbf{i}_2 + (a_1b_2 - a_2b_1)\mathbf{i}_3\} = 0.$

7. $4\pi a$.

8. $-\dfrac{1}{2}\left(\log 2 - \dfrac{\pi}{4}\right), -\dfrac{1}{2}\left(\log 2 + \dfrac{\pi}{4}\right), \dfrac{\pi\sqrt{2}}{8}.$

9. $\pi/4;\ \pi/3;\ \left(\dfrac{\pi}{2} - \cos^{-1}\dfrac{13}{10\sqrt{2}}\right).$

Index